PHYSICS THROUGH THE 1990s

An Overview

Physics Survey Committee
Board on Physics and Astronomy
Commission on Physical Sciences,
Mathematics, and Resources
National Research Council

NATIONAL ACADEMY PRESS
Washington, D.C.   1986

NATIONAL ACADEMY PRESS 2101 Constitution Avenue, NW Washington, DC 20418

NOTICE: The project that is the subject of this report was approved by the Governing Board of the National Research Council, whose members are drawn from the councils of the National Academy of Sciences, the National Academy of Engineering, and the Institute of Medicine. The members of the committee responsible for the report were chosen for their special competences and with regard for appropriate balance.

This report has been reviewed by a group other than the authors according to procedures approved by a Report Review Committee consisting of members of the National Academy of Sciences, the National Academy of Engineering, and the Institute of Medicine.

The National Research Council was established by the National Academy of Sciences in 1916 to associate the broad community of science and technology with the Academy's purposes of furthering knowledge and of advising the federal government. The Council operates in accordance with general policies determined by the Academy under the authority of its congressional charter of 1863, which establishes the Academy as a private, nonprofit, self- governing membership corporation. The Council has become the principal operating agency of both the National Academy of Sciences and the National Academy of Engineering in the conduct of their services to the government, the public, and the scientific and engineering communities. It is administered jointly by both Academies and the Institute of Medicine. The National Academy of Engineering and the Institute of Medicine were established in 1964 and 1970, respectively, under the charter of the National Academy of Sciences.

Library of Congress Cataloging-in-Publication Data
Main entry under title:

Physics through the 1990s: An overview

(Physics through the 1990s)
Includes index.
1. Physics. I. National Research Council (U.S.).
Physics Survey Committee. II. Series.
QC21.2.P4794   1986   530'.072073   85-30999
ISBN 0-309-03578-3, soft cover
ISBN 0-309-03581-3, hard cover

Printed in the United States of America

PHYSICS SURVEY COMMITTEE

WILLIAM F. BRINKMAN, Sandia National Laboratories, *Chairman*
JOSEPH CERNY, University of California, Berkeley, and Lawrence Berkeley Laboratory
RONALD C. DAVIDSON, Massachusetts Institute of Technology
JOHN M. DAWSON, University of California, Los Angeles
MILDRED S. DRESSELHAUS, Massachusetts Institute of Technology
VAL L. FITCH, Princeton University
PAUL A. FLEURY, AT&T Bell Laboratories
WILLIAM A. FOWLER, W. K. Kellogg Radiation Laboratory
THEODOR W. HÄNSCH, Stanford University
VINCENT JACCARINO, University of California, Santa Barbara
DANIEL KLEPPNER, Massachusetts Institute of Technology
ALEXEI A. MARADUDIN, University of California, Irvine
PETER D. MACD. PARKER, Yale University
MARTIN L. PERL, Stanford University
WATT W. WEBB, Cornell University
DAVID T. WILKINSON, Princeton University

DONALD C. SHAPERO, *Staff Director*
ROBERT L. RIEMER, *Staff Officer*
CHARLES K. REED, *Consultant*

BOARD ON PHYSICS AND ASTRONOMY

HANS FRAUENFELDER, University of Illinois, *Chairman*
FELIX H. BOEHM, California Institute of Technology
RICHARD G. BREWER, IBM San Jose Research Laboratory
DEAN E. EASTMAN, IBM T.J. Watson Research Center
JAMES E. GUNN, Princeton University
LEO P. KADANOFF, The University of Chicago
W. CARL LINEBERGER, University of Colorado
NORMAN F. RAMSEY, Harvard University
MORTON S. ROBERTS, National Radio Astronomy Observatory
MARSHALL N. ROSENBLUTH, University of Texas at Austin
WILLIAM P. SLICHTER, AT&T Bell Laboratories
SAM B. TREIMAN, Princeton University

DONALD C. SHAPERO, *Staff Director*
ROBERT L. RIEMER, *Staff Officer*
HELENE PATTERSON, *Staff Assistant*
SUSAN WYATT, *Staff Assistant*

COMMISSION ON PHYSICAL SCIENCES, MATHEMATICS, AND RESOURCES

HERBERT FRIEDMAN, National Research Council, *Chairman*
THOMAS D. BARROW, Standard Oil Company (Retired)
ELKAN R. BLOUT, Harvard Medical School
WILLIAM BROWDER, Princeton University
BERNARD F. BURKE, Massachusetts Institute of Technology
GEORGE F. CARRIER, Harvard University
CHARLES L. DRAKE, Dartmouth College
MILDRED S. DRESSELHAUS, Massachusetts Institute of Technology
JOSEPH L. FISHER, Office of the Governor, Commonwealth of Virginia
JAMES C. FLETCHER, University of Pittsburgh
WILLIAM A. FOWLER, California Institute of Technology
GERHART FRIEDLANDER, Brookhaven National Laboratory
EDWARD D. GOLDBERG, Scripps Institution of Oceanography
MARY L. GOOD, Signal Research Center
J. ROSS MACDONALD, University of North Carolina
THOMAS F. MALONE, Saint Joseph College
CHARLES J. MANKIN, Oklahoma Geological Survey
PERRY L. MCCARTY, Stanford University
WILLIAM D. PHILLIPS, Mallinckrodt, Inc.
ROBERT E. SIEVERS, University of Colorado
JOHN D. SPENGLER, Harvard School of Public Health
GEORGE W. WETHERILL, Carnegie Institution of Washington

RAPHAEL G. KASPER, *Executive Director*
LAWRENCE E. MCCRAY, *Associate Executive Director*

Contents

FOREWORD xiii

PREFACE xv

ACKNOWLEDGMENTS xvi

SUMMARY 1
Background, 1
Physics and the Nation, 2
Universities and Small-Group Research, 3
Large Facilities and Major Programs, 3
Supporting Physics Research, 4
Manpower, 4
International Position of U.S. Physics, 5

1   PHYSICS AND SOCIETY 6

2   PROGRESS IN PHYSICS 11
Introduction, 11
   Elementary-Particle Physics, 11
   Nuclear Physics, 12

Condensed-Matter Physics, 13
Atomic, Molecular, and Optical Physics, 13
Plasma Physics, 14
Cosmology, Gravitation, and Cosmic Rays, 14
Interfaces and Applications, 15
The Unity of Physics, 15
Progress in Particle Physics, 18
Quarks and Leptons as Elementary Particles, 18
Unification of the Forces of Nature, 20
Progress in Nuclear Physics, 21
Progress in Condensed-Matter Physics, 24
Surfaces, Interfaces, and Artificially Structured Materials, 24
Phase Transitions and Disordered Systems, 25
Progress in Atomic, Molecular, and Optical Physics, 28
Progress in Plasma and Fluid Physics, 31
Progress in Plasma Physics, 31
Fusion, 32
Space Plasmas, 34
Fluid Physics, 35
Progress in Gravitation, Cosmology, and Cosmic-Ray Physics, 35
Gravitational Physics, 35
Cosmology, 37
Cosmic-Ray Physics, 38
Interfaces and Applications, 39
Interface Activities, 39
Chemistry, 39; Biophysics, 40; Geophysics, 40; Materials Science, 41
Applications, 42
Energy and the Environment, 42; Medicine, 42; National Security, 43; Industry, 43

3   MAINTAINING EXCELLENCE. 44
The Funding Process, 45
Educating the Next Generation of Physicists, 46
Primary and Secondary Education, 46
Undergraduate Education, 47

Education at the Graduate Level, 48
Research in Small Groups, 49
Large Facilities and Major Programs, 53
  Elementary-Particle Physics, 58
    The Superconducting Super Collider, 58; Extensions of the
    Capabilities of Existing Accelerators, 60; Support of
    Existing and Extended Facilities, 61
  Nuclear Physics, 61
    The Continuous Electron Beam Accelerator Facility,
    62; The Relativistic Nuclear Collider, 62; Extensions of
    Existing Facilities, 62
  Condensed-Matter Physics, 63
    Synchrotron Radiation Facilities, 63; Neutron Facilities,
    64; High Magnetic Fields, 65
  Plasma Physics, 65
    Magnetic Fusion Research, 65; Inertial Fusion
    Research, 66
  Space and Astrophysical Plasmas, 67
  Gravitation, Cosmology, and Cosmic-Ray
    Physics, 67
    Search for Gravitational Radiation, 68; Relativity Gyroscope Experiment, 68; Vigorous Space Program in
    Astrophysics, 68; Long-Duration Cosmic-Ray
    Experiments, 69; Ground-Based Cosmic Rays, 69;
    Neutrino Astronomy, 69
Manpower and Excellence, 69
Policy Issues Connected with Maintaining
    Excellence, 70
  Role of Industry and Mission Agencies in Basic
    Research, 71
  Freedom of International Communication and
    Exchange, 71
Computation and Data Bases, 72
  Computers, 72
  Data Bases, 73

SUPPLEMENT

1 INTERNATIONAL ASPECTS OF PHYSICS: THE U.S.
  POSITION IN THE WORLD COMMUNITY 75
  Expenditures for Scientific Research in the
    United States and Abroad, 76

General Trends, 76
Trends in Specific Areas of Physics, 79
The U.S. Position in Basic Physics Research, 81
International Competition and Cooperation, 85
  Increased Internationalization of the Physics
    Community, 85
  Scale and Costs, 85
  Avoiding Duplication, 86
  Maintaining Breadth and Depth in Forefront
    Areas, 86
  Freedom for Scientists and the Free Flow of
    Information, 87
  Education of Foreign Physicists in the
    United States, 87
Summary, 90

SUPPLEMENT

2   EDUCATION AND SUPPLY OF PHYSICISTS . . . 91
    Producing Trained Young Physicists—A Historical
      Overview, 92
    Enrollments and Degrees: The Prolonged
      Decline, 95
      U.S. and Foreign Composition, 95
      Women and Minorities, 95
      Declining Enrollments in Physics Subfields, 98
    Retention of Physics Degree Holders—Mobility, 98
    An Aging Community, 99
    Changing Patterns of Employment, 101
    Projections, 102
      Demand Projections, 103
        Academe, 103; Demand Scenarios—Universities, 105;
        Demand Scenarios—4-Year Colleges, 105; Demand
        Scenarios—Industrial and Other Nonacademic Sectors, 106
      Supply Projections, 108
        Physics Ph.D. Production, 109; Supply of Physics Ph.D.s:
        1981-2001, 111
      The Demand-Supply Balance, 114
    Conclusion, 114

SUPPLEMENT
3 ORGANIZATION AND SUPPORT OF PHYSICS 115
   The Diversity of Institutions for Research in
      Physics, 115
      Major Facilities and National Laboratories, 116
      University Research, 117
      Industrial Research, 117
   The Complementary Roles of Our Research
      Institutions, 118
   Funding Support for Physics Research, 119
   Organization and Decision Making, 135

ABBREVIATIONS AND ACRONYMS 139

GLOSSARY OF PHYSICAL TERMS 142

APPENDIX A: PANEL MEMBERS 157

INDEX 161

Foreword

PHYSICS THROUGH THE 1990s is an eight-volume survey of physics that documents the extraordinary accomplishments of physicists over the decade since the last such survey was completed. The survey also assesses opportunities for the next decade and addresses some of the obstacles that must be overcome if those opportunities are to be realized.

The breadth and diversity of physics as portrayed in these volumes is truly breathtaking. Physics examines phenomena across an enormous range, from the subatomic to the cosmic. It is concerned with fundamental questions about the origins of the universe and the structure of matter that have applications in virtually all human endeavors. Progress in physics has touched almost every science and every aspect of industry and technology with new ideas, new instruments and techniques, and new applications.

Our whole picture of the nature of space and time and the elementary building blocks of matter is undergoing revolutionary change; the pace of revelations has accelerated with each passing year so that the productivity and accomplishments of physics have outstripped the most ambitious hopes of physicists of a decade ago.

The development of new physics-based imaging technologies is ushering in a new era in medicine in which physiological functions can be mapped out with precision and in exquisite detail without even touching the patient. New techniques in solid-state physics, in which semiconductor devices can be tailored on the atomic scale, promise to introduce a new generation of high-speed devices that will form the basis for smaller, faster, and less costly computers of the future.

It is unfortunate, in a sense, that these unique advances of physics should be occurring at a time of budgetary constraint that will affect the nation's scientific enterprise in many ways. But this conjunction of events does not detract from the value of the survey. Decisions must be made about the direction of scientific programs even in difficult times; what is more, the budgetary problems will eventually be overcome. In any case, those who direct our scientific research efforts will always require a clear picture of the state of the fields of science.

I commend this overview volume to you as a presentation of the full panoply of scientific accomplishments and opportunities of physics. The conclusions and recommendations of this volume merit thoughtful consideration by decision makers concerned with physics and its application in academe, industry, and the federal government. The story it tells, fascinating in itself, cannot fail to fill us with expectations of even more spectacular accomplishment in the coming decades.

FRANK PRESS, *Chairman*
National Research Council

Preface

The Physics Survey Committee took as its task to carry out a research assessment of the major fields of physics. The purpose of the assessment is to review the developments that have taken place since the last survey and to highlight research opportunities. This task is one of considerable scope, as the eight volumes that constitute the Physics Survey attest. In addition to this *Overview*, the volumes include *Atomic, Molecular, and Optical Physics*; *Condensed-Matter Physics*; *Elementary-Particle Physics*; *Gravitation, Cosmology, and Cosmic-Ray Physics*; *Nuclear Physics*; *Plasmas and Fluids*; and *Scientific Interfaces and Technological Applications*. These volumes document a physics enterprise that is vital, creative, and productive.

A number of critical questions emerged in the course of the assessment effort, including the following: What are the areas of physics that showed the greatest progress over the past decade? What are the problems of educating the next generation of physicists? Will physics continue to provide the scientifically and technologically trained manpower required by our society? What is the U.S. position in the world physics community? Does our scientific support system still support excellence, in small projects and large? These questions are addressed in the present volume, *Physics Through the 1990s: An Overview*.

WILLIAM F. BRINKMAN, *Chairman*
Physics Survey Committee

Acknowledgments

The Physics Survey Committee acknowledges the contributions of the many groups that helped to complete the survey: the panel members (listed in Appendix A of this report); federal agencies listed below for assistance in developing data for Supplement 3; Beverly Fearn Porter and Roman Czujko of the American Institute of Physics for their help in preparing Supplement 2 on education and manpower; the National Research Council's Board on Physics and Astronomy and its staff for advice and assistance in carrying out the study; and Jacqueline Boraks for editing and preparing the final manuscripts for publication.

The Board on Physics and Astronomy is pleased to acknowledge generous support for the Physics Survey from the Department of Energy, the National Science Foundation, the Department of Defense, the National Aeronautics and Space Administration, the Department of Commerce, the American Physical Society, Coherent (Laser Products Division), General Electric Company, General Motors Foundation, and International Business Machines Corporation.

Summary

BACKGROUND

Physics encompasses the broad search for basic knowledge and the search for technology applicable to urgent societal problems; the research is pursued in many different institutional settings including national laboratories, government laboratories, industrial research centers, and universities. A single volume that attempted to portray both the full scope of these activities and the many roles of physics on the national scene would be unmanageable. Consequently, the main body of the Physics Survey is presented in a series of reports prepared by panels of the Survey Committee—one report on each major subfield of physics and one on scientific interfaces and technological applications. These seven reports, together with this *Overview*, collectively constitute the Physics Survey, *Physics Through the 1990s*. The subtitles are

- *An Overview*
- *Atomic, Molecular, and Optical Physics*
- *Condensed-Matter Physics*
- *Elementary-Particle Physics*
- *Gravitation, Cosmology, and Cosmic-Ray Physics*
- *Nuclear Physics*
- *Plasmas and Fluids*
- *Scientific Interfaces and Technological Applications*

This *Overview* summarizes the findings of the panels (the panels are listed in Appendix A) and addresses issues that broadly concern physics. The role of physics in society is discussed in Chapter 1. Highlights of the progress and the opportunities in each subfield are presented in Chapter 2. Future needs and recommended action are described in Chapter 3. In addition, there are three supplements on issues that cut across the individual fields: international aspects of physics, education and supply of physicists, and organization and support of physics.

PHYSICS AND THE NATION

We are witness to one of the most exciting times in physics; major advances are to be found in every field. In *particle physics* theories of the electromagnetic and weak forces have been unified with one theory that explains electromagnetic and radioactive decay phenomena in a consistent manner. The particles that carry the weak interaction that were predicted by the unified theory have been observed. The multitude of subnuclear particles that have been generated by high-energy particle accelerators can now be described in terms of small families of elementary particles called quarks and leptons. The forces that hold quarks together are beginning to be explained by particles called gluons. In *nuclear physics*, after decades of work in which studies of nuclear systems succeeded in revealing the behavior of the particles (called nucleons) that make up atomic nuclei, we now have the possibility of creating an entirely new state of matter—one in which the constituents of the nucleons themselves (quarks and gluons) emerge to form a plasma.

In *plasma and fluid physics*, magnetically confined plasmas have been created at densities and temperatures that approach the conditions required to produce a fusion reaction in which nuclei merge with a release of energy. These conditions have also been approached using lasers, with which physicists have compressed pellets to many times their liquid density. Studies of fluids and plasmas have led to dramatic progress in understanding turbulence and chaos. The revolution in the field of *atomic, molecular, and optical physics* caused by the laser continues. Spectroscopic accuracy has been enormously enhanced, and new atomic and molecular species have been discovered.

In *condensed-matter physics*, techniques developed to explore the nature of phase transitions (changes in the state of matter from, for example, liquid to solid) have helped to elucidate disordered systems. States of matter found nowhere in nature have been created artificially, and unexpected phenomena such as the quantized Hall effect have been discovered. In *cosmology and gravitation* new observational techniques are transforming the study of the universe. The confluence of cosmology with elementary-particle physics and condensed-matter physics has created a new picture of the origin and development of the universe. Experimental gravitational physics is emerging as a new discipline.

Viewed collectively, these discoveries in physics are among the greatest achievements of our time. The impact of physics extends far beyond the satisfaction of man's desire to understand nature, for physics is a central discipline that contributes theoretical concepts and experimental techniques to all the other natural sciences, to technology, and to medicine. Physics is a vital component of such national programs as energy development, environmental improvement, and security. Discoveries in physics have generated entire industries such as microelectronics and optical technology. By helping the nation to maintain technological leadership, physics constitutes a driving force in our economy.

UNIVERSITIES AND SMALL-GROUP RESEARCH

Concern is felt in all of physics for the health of the nation's university research institutions that will educate the next generation of physicists. Retirements from physics department faculties will begin to occur at an increasing rate starting in the early 1990s. To meet the need for faculty replacements, steps should be taken to ensure the continued ability of universities to attract highly qualified young physicists to work in an academic setting. This need is particularly acute in fields where research is carried out by small groups. Such groups make an exceptionally strong contribution to educating new physicists. To enhance the attractiveness of academic research, the difficulty in obtaining modern instrumentation in university research laboratories and the difficulty in obtaining support for research groups must be addressed. The resources required represent only a tiny fraction of the nation's total research and development expenditures. For our universities to maintain preeminence in physics and train graduate students at the highest level, the support for university-based research must be increased to ensure that it can compete at the forefront of physics.

LARGE FACILITIES AND MAJOR PROGRAMS

Large national facilities and major programs in physics are essential for forefront research. To assist the Congress and funding agencies that are responsible for the planning of these facilities and programs, the panel reports on each of the relevant subfields describe priorities in detail, justifying the proposed facilities and programs and explaining the process by which recommendations were formulated. These findings are summarized in Chapter 3 of this *Overview* along with a brief discussion of priority setting in physics.

SUPPORTING PHYSICS RESEARCH

The support of basic research in physics and all the physical sciences, which is essential to ensure the future technical leadership of this country, is approximately 2.5 percent of the total research and development expenditures in the United States. The fraction spent on basic research is not large compared with the total research and development expenditures, and a healthy physics research enterprise is well within the nation's means.

MANPOWER

The production of Ph.D. physicists has remained stable for more than a decade. However, the proportion of foreign-born students has steadily grown; today about 40 percent of our entering graduate students are from abroad. Young physicists have increasingly found work in industry; approximately one third of the new Ph.D.s leave physics research. At present, a balance exists between the supply of and demand for scientific manpower in physics, but it is precarious and could be upset by a change in the career patterns of foreign-born scientists or the creation of large government programs. Starting in the 1990s, however, the faculty retirement rate will begin to increase in universities and colleges throughout the nation, and we can predict a shortage of qualified applicants for academic positions. Such a shortage would have most serious consequences not only for the quality of undergraduate and graduate education but also for the quality of basic research in the universities. Further, because a majority of those trained as physicists are now engaged in applications or engineering, the increasing technological focus of the U.S. economy may increase demand for physics graduates even more rapidly.

In most of the sciences in the United States, the number of male scientists has decreased, but the number of female scientists has increased to fill the gap. The number of female physicists, however, remains small; steps should be taken to realize the potential of this untapped resource.

INTERNATIONAL POSITION OF U.S. PHYSICS

For 40 years, the United States has been the world leader in physics research, but the situation is changing rapidly. During the past decade, both the Western European nations and Japan have fully recovered from World War II and have reassumed an aggressive role in science. In many areas where previously we were clearly ahead, these nations are now fully competitive. Their re-emergence in the field of physics benefits science as a whole, but the United States can and must retain a competitive edge. Without it an essential factor in maintaining our economic well-being will be lost.

1

Physics and Society

Physics is the science of the most fundamental aspects of nature. The realms of physics span distances from the subnuclear world of elementary particles to the whole of the cosmos, and times from less than a billionth of a trillionth of a second to the age of the universe. To study phenomena across these epochal scales, to devise experimental tools that provide ever more powerful means for viewing nature, and to create theories that allow us to comprehend what has been seen—these are the goals and the achievements of physics. In deepening our view of nature, physics has profoundly affected our view of mankind because the underlying assumption of physics—that there is order in the natural world and that the human mind can understand that order—permeates modern thought. By generating new technologies and by contributing to neighboring sciences, physics has helped to transform our daily lives, permitting a comfort and freedom of action that make it difficult to comprehend that little more than a century ago, even in the technically advanced nations, most people devoted most of their energy to securing food and shelter.

Physics has done much to mold the shape of modern society. The search to understand elementary phenomena has led to expanded views of all nature and to miraculous inventions. The path of this search is unpredictable, but along it the history of physics offers many examples. To cite one of these, starting in the eighteenth and nineteenth centuries, scientists like Cavendish, Franklin, Ampère, and

Faraday carried out painstaking experiments on electrical and magnetic phenomena. Their observations provided the foundation for James Clerk Maxwell's electromagnetic theory. Maxwell discovered that light is a natural manifestation of familiar electrical and magnetic forces. To find a connection between light and everyday forces was a triumph of modern thought. From Maxwell's work we have gained a deep understanding of electromagnetic phenomena from waves to plasmas and the answers to questions ranging from why air is transparent to how radiative energy transport determines solar and stellar equilibrium. Electromagnetic theory underlies the invention of radio, television, and radar and makes possible the creation of our vast industrial power networks and modern communications systems. In fact, it is difficult to cite instances of modern life that have not been touched in some way by Maxwell's discovery.

The creation of quantum mechanics in the 1920s provides a second example of the unpredictable path by which new knowledge in physics can shape society. Based on studies of the properties of matter, the spectra of atoms, and the motions of charged particles, quantum mechanics provided an extraordinary new framework for portraying physical reality. Quantum mechanics revolutionized our most fundamental concepts of measurement and paved the way to understanding the structure of atoms, molecules, and solids. It is now recognized that quantum mechanics is basic not only to physics but to chemistry, biology, and many of the other sciences. Beyond this, quantum mechanics has led to the creation of new industries, such as semiconductors and optical communications, and has opened new paths of technology through the creation of exotic materials and devices like the laser.

The discoveries of electromagnetic theory and quantum mechanics are now part of history, but seminal advances in physics continue. A few decades ago no one realized that cosmology and astrophysics were on the threshold of a golden age and that radioastronomy, x-ray astronomy, and other new techniques were about to yield pictures of the universe that would make our previous views seem blurred and shadowy images. A marvelous scenario of the origin of the universe and its eventual fate is now being constructed from these pictures. The new theories cannot help but eventually affect our total vision of our role in nature and deepen our understanding and appreciation of life.

Another advance can be traced to the study of electronic materials known as semiconductors. The discovery in 1947 by Shockley, Bardeen, and Brattain of the transistor effect paved the way for the computer revolution that is taking place around us today. Nobody can know how society will ultimately be transformed by this revolution, but the advances have been so rapid that the image of a savings bank with clerks patiently entering transactions by hand, without the benefit of automatic data processing, seems almost as remote as a candlelit counting house in a novel by Dickens.

Before World War II, physics was essentially a European activity, but by the war's end, the center of physics had moved to the United States. The influx of European physicists, a generation of outstanding young American physicists, and the heritage of intense cooperation between science and government that grew out of the nation's efforts to develop radar and the atomic bomb presented the United States with a scientific community of unsurpassed quality. Discoveries by this community since World War II rank among the greatest achievements of physics. They include the creation of quantum electrodynamics, the theory of superconductivity, the discovery of remnants from the primordial explosion at the birth of the universe, and the invention of the transistor and the laser.

The role of physics in the United States today is complex. Curiosity and the basic need for understanding remain the intellectual driving force for physics, but physics also affects society broadly through its interactions with all the natural sciences, with technology and engineering, and with medicine. To strengthen these, physics provides conceptual tools, experimental techniques, and new materials. Much of our advanced technology can be traced directly to basic research in physics. Optical communications and laser-assisted manufacturing exemplify technologies that will play a vital role in helping the nation to retain its industrial competitiveness in the years to come.

Biophysics, molecular biology, and physiology—three sciences that closely underlie medicine—all use concepts and experimental techniques from physics. Nuclear medicine, radiation therapy, x-ray tomography, and laser surgery are but a few of physics' many contributions to medical diagnosis and therapy. One of the most recent diagnostic techniques, magnetic resonance imaging (MRI), allows doctors to peer into the human body as clearly as they can view its surface (Figure 1.1). MRI is nonperturbing, noninvasive, and free of any known side effects. Its role in medical diagnosis is expected to be revolutionary, comparable in effect to the discovery of x rays. The creation of MRI required microcomputers—whose origin depended on the invention of the transistor—and superconducting magnets—which were created from research in low-temperature physics. The basic principles of MRI, however, were discovered in pioneering research by Purcell and Bloch, who were simply curious about how nuclei magnet-

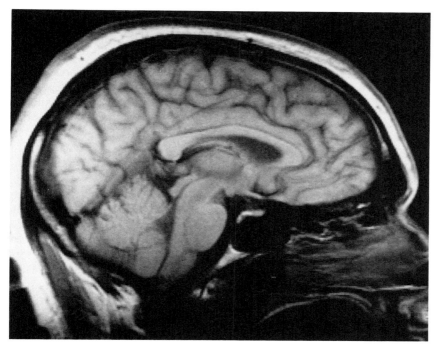

FIGURE 1.1 Proton magnetic resonance image showing a sagittal slice of cranial anatomy demonstrating an enlarged pituitary gland. The image was obtained on a Technicare 1.5-tesla Teslacon magnetic resonance system at the Cleveland Clinic using a spin-echo pulse technique. The phenomena of nuclear magnetic resonance, solid-state microelectronics, and superconducting materials for high field magnets, which combine to make the medical magnetic resonance imager, are all products of physics research of the past four decades.

ically interact with matter. The benefits now flowing from their work illustrate the large returns in human well-being that may come from basic research by scientists with the freedom and the resources to pursue the search for fundamental understanding.

Physics has given mankind the power to make life better or to destroy it. How to use this power wisely has become society's most urgent challenge. To meet it successfully will require a public educated in the underlying science and a political leadership well informed about scientific issues and technical options. Physicists must play an essential role in advising and counseling.

The United States is a leader in the community of nations. Our economic vitality and our national security system are visible signs of our strength, but our authority stems ultimately from the political and social ideals that animate us and from our cultural and intellectual accomplishments. Our achievements in physics are respected by people everywhere; they help the United States to fulfill its role as a world leader.

Physics in the United States has become a federal responsibility. Universities, industry, and private institutions all participate in physics research, but the federal government has assumed major responsibility for its support. This assumption reflects the essential role that physics plays in generating new technologies and in maintaining our national defense. It also reflects a widespread public interest in science, an excitement over new discoveries, and a national pride in accomplishment.

Advanced technology, in this age of robotics and the information explosion, is a major driving force behind economic growth in the developed countries. Looking to the future, the United States must be able to meet the challenge of creating the new technology needed to sustain growth. And the other nations of the world must be able to meet the challenges posed by the growth of population, the pressure on nonrenewable resources, and the increasing burden on the environment. Many nations will look to the United States for help in providing the new technologies required to solve their problems. If the world is to live in peace, we must meet these challenges. Basic science is the driving force of the new technology; the role that physics can play is critical to the future of mankind.

2

Progress in Physics

INTRODUCTION

We live in one of the most productive eras in the history of physics. This chapter highlights some advances and opportunities in physics that have been culled from the multitude of achievements reported in the accompanying volumes of the Physics Survey. Our discussion begins with a brief, nonspecialized summary of the highlights and a description of the unifying principles that join the different subfields of physics.

Elementary-Particle Physics

Elementary-particle physics, the science of the ultimate constituents of matter and their interactions, has undergone a remarkable development during the past two decades. A host of experimental observations made possible by the current generation of particle accelerators and the accompanying rapid convergence of theoretical ideas have led to a radically new and simple picture of nature. All matter in its infinite diversity has been found to be composed of a few basic constituents called quarks and leptons, which are structureless and indivisible at current limits of resolution. Great progress has also been made in understanding the character of fundamental forces that govern natural phenomena. The weak and electromagnetic interactions have been unified in a theory whose predictions have been verified by many experiments, culminating in the 1983 discovery of the W and Z particles, the mediators of the weak interaction. The similarity among quarks and leptons and the mathematical resemblance among the theories of fundamental interactions spur bold attempts at unification in which all the fundamental forces are seen as different manifestations of a single underlying symmetry, a symmetry that is partially hidden. The new synthesis raises deep questions about family patterns of quarks and leptons and the origin of particle masses and invites speculation about the eventual compositeness of quarks and leptons themselves. These concerns motivate a broad program of experimentation at higher energies (and, equivalently, on shorter scales of distance and time) to test the emerging standard model and to uncover clues leading to more complete understanding.

Nuclear Physics

During the past decades, the building blocks of nuclei were thought to be protons and neutrons bound together by mesons. Today we know that protons and neutrons are made of quarks and that the forces between the quarks are created by particles called gluons. The new concept is based on theoretical advances in particle physics, but recent experimental work has demonstrated the importance of this description for nuclei also. The basic questions facing nuclear physics today involve detailed exploration of the quark structure in nucleons and nuclei and the strong many-body forces that confine quarks and gluons. Finding the answers represents an exciting frontier that may lead to more basic understanding of the strong forces and of nuclear structure and dynamics.

Heavy ions have been used to probe nuclear dynamics under extreme conditions and to create new elements. Theoretical investigation predicts the existence of a quark-gluon plasma similar to that which may have existed in the earliest moments of our universe. Studies are being conducted on new systems made in the laboratory called hypernuclei, in which a quark has been replaced by a strange quark. The use of high-energy electron scattering from nuclei is now revealing unprecedented levels of detail of nuclear structure, probing the electroweak interactions between nucleons and their underlying quark components. Finally, nuclear science continues to have great impact on our understanding of fundamental symmetries in physics, while also playing an ever increasing role in astrophysics and cosmology.

Condensed-Matter Physics

Condensed-matter physics has continued its historic role as a major source of new concepts in fundamental science from the explanation of the behavior of neutron stars, through advances in our understanding of semiconductors, superconductors, and magnetisms, to prediction and discovery of a new state of matter, the superfluid phase of liquid helium atoms of mass 3. This is the area of physics that most directly fuels advances in technology, from jet engines to computers. Conceptual advances abound, increasingly stimulated by the creation of totally new substances not found in nature. Some of these substances are produced by novel experimental techniques, such as exceedingly rapid cooling of liquids to the solid state or controlled deposition of atoms layer by layer. Others are produced by more conventional means on the basis of new theoretical hypotheses made possible by deepening theoretical understanding. Areas of great activity include studies of systems of one or two dimensions, studies of phenomena at surfaces, the role of interfaces between different materials, disordered systems, surprising new forms of ordered systems, the onset of turbulence in liquids, and the possibility of new forms of superconductivity.

Atomic, Molecular, and Optical Physics

This field has been revolutionized by the laser and modern optics. New atomic and molecular species have been created using laser light; spectroscopic resolution has been increased more than a millionfold. Lasers have made it possible to watch atoms as they collide and chemical reactions as they take place. Lasers are now being used to generate femtosecond light pulses and coherent soft x rays and to cool atoms to the submillikelvin regime. Optical-frequency counting methods using laser light have become so precise that the meter is no longer defined in terms of the wavelength of light but as the distance light travels in a given time interval. Particle-trap techniques have led to ultraprecise studies of quantum electrodynamics and mass spectra; they have made it possible to study plasma liquids and to create new kinds of atomic clocks. Today's research opportunities include ultrasensitive tests of the properties of space and the symmetries in nature, studies of relativistic many-body theory and quantum electrodynamics in heavy ions by advanced x-ray spectroscopy, new experimental and theoretical approaches to the structure and interaction of atoms and molecules, and the creation of nonlinear optical techniques and new light sources.

Plasma Physics

Most of the visible matter in the universe is made of plasmas—neutral gases composed of positive ions and unbounded electrons. Our understanding of stars, stellar winds, planetary magnetospheres, and galaxies is being spurred by advances in plasma physics. Spacecraft have probed the magnetospheres of the planets from Mercury to Saturn and soon will reach Uranus. The solar wind has been monitored by many spacecraft, from inside the orbit of Mercury to beyond Pluto. We may have data on galactic plasma beyond the influence of the Sun before the year 2000. The Earth's magnetosphere has been measured in great detail, and we are beginning to understand the complex phenomena seen there—its weather, so to speak.

On Earth, our mastery of high-temperature plasmas has advanced remarkably. Today, one plasma-confinement approach is expected soon to achieve breakeven conditions for controlled fusion employing reactions similar to those that power the Sun and stars. Our understanding of plasmas is having an impact on physics in many other ways. One example is the recent generation of electric fields of tens of millions of volts per centimeter in plasmas by the excitation of an electron-plasma oscillation. It is expected that the technique can be extended to give fields hundreds of times larger, i.e., as large as the electric field that holds electrons in atoms. The possibility of using these fields to accelerate particles to high energy is being explored. Another example is the development of the free-electron laser, which can generate coherent radiation from microwaves to the ultraviolet.

Cosmology, Gravitation, and Cosmic Rays

The study of the universe is being transformed by new eyes, such as x-ray and infrared telescopes in space and very-large-array radio telescopes on the ground. A vivid history of the universe has emerged, starting with a primordial explosion—the big bang—about 15 billion years ago. Recent discoveries from elementary-particle theory are offering possible solutions to some of the profound questions in cosmology (for instance, why the universe appears to be so uniform, and why there is so much matter relative to antimatter). An intense search is under way for dark matter in the universe; such matter may dominate important processes such as the formation of galaxies and the ultimate fate of the universe. Will the universe expand forever or collapse to start anew in yet another primordial explosion? Fundamental theoretical advances in gravitational physics are leading to a better understanding of black holes and quantization of gravity, and a prediction of Einstein's General Theory of Relativity has been verified to 2 parts in 1000, using the Viking spacecraft on Mars. Basic theoretical and experimental research has now prepared the way for major efforts to detect gravitational radiation. This dramatic advance could take place in the next decade and open an important new window on the universe. Space-based studies of the enigmatic cosmic rays suggest that they originate in interstellar space, while ground-based instruments have discovered localized sources of gamma rays with 10^{15} eV of energy.

Interfaces and Applications

Research at the boundaries between physics and neighboring areas such as chemistry, biology, materials science, and mathematics has blossomed with new ideas and new approaches. High-vacuum and surface science, the transition from orderly to chaotic motion, polymer and macromolecular structure, the origins of biological processes, microscopic control of structure and function in liquids and solids—these are but a few examples of physics' rapidly emerging interdisciplinary advances, which are enriching all of science.

The applications of physics are broad and affect virtually every area of society (see Figure 2.1). Our national programs in energy, the environment, medicine, and security depend critically on physics. Our industrial posture is linked to the flow of discoveries from physics that can lead to the creation of industries such as microelectronics and optical technology.

THE UNITY OF PHYSICS

The highlights in the preceding section hardly begin to portray current advances in physics and their effects on science and society. The following sections of this chapter provide a somewhat more detailed account; but for a comprehensive picture of today's research and tomorrow's opportunities, the reader is referred to the seven panel reports of the Physics Survey that accompany this overview volume.

The scope of physics is so broad and its styles of research so diverse that it is easy to lose sight of the underlying unity that joins even the most disparate activities into a common enterprise. This unity is a fundamental source of the strength and vitality of physics: to under-

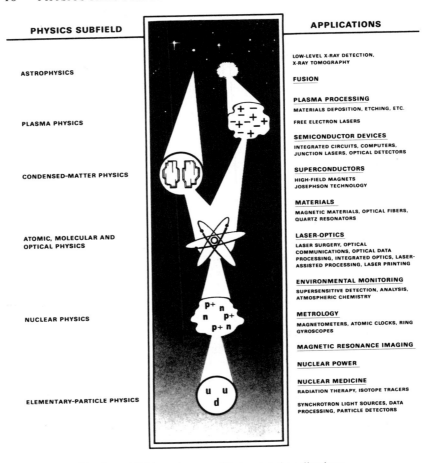

FIGURE 2.1 Physics subfields and some of their related applications.

stand physics it must be appreciated. We present some examples below.

A dramatic confluence of ideas from three diverse subfields illustrates the unexpected connections in physics that suddenly occur. Insights from particle physics based on the quark-gluon model (the modern theory of the structure of protons, neutrons, and other subnuclear particles) have been combined with contemporary ideas from condensed-matter theory to portray the evolution of the universe in the earliest stages of the primordial explosion—the big bang. This synthesis of thought allows us to understand important features of the universe that we can observe today as consequences of elementary ideas about the structure and organization of matter.

Physicists are drawing on techniques from nuclear, condensed-matter, and atomic physics to address another cosmological problem—that of the missing mass in the universe. A critical question is whether a particle known as the neutrino has a finite mass or whether it is massless like the photon. The most sensitive laboratory experiments appear to show that the neutrino's mass must be less than one ten-thousandth the mass of the electron, but the question of the neutrino's mass is not yet settled. If it is large enough, general relativity predicts that the universe will not expand forever but that it will eventually collapse.

A further example of links between diverse areas is the renewed interest in the relation between regular and chaotic motion. Abrupt transitions from regular to chaotic behavior have been discovered in electrical, acoustical, and optical systems, in fluid flow, in chemical reactions, and in the behavior of simple differential equations. Cardiac arrest due to fibrillation of the heart is believed to be due to such an effect. Recognition of the universal nature of such transitions gives hope for understanding chaotic motion in more complex systems and, in particular, of understanding turbulence. This line of research can be expected to have a deep influence on many areas of science and on problems ranging from aircraft and ship design to weather forecasting.

The subfields of physics are joined by technical as well as conceptual bonds. Lasers, for example, have had a dramatic effect on science and technology. They have revolutionized spectroscopy by enormously increasing its sensitivity and precision and have opened the way to the creation of new types of atomic and molecular species. Femtosecond (a millionth of a billionth of a second) laser pulses make it possible to take "snapshots" of chemical reactions; nonlinear spectroscopy makes it possible to study reactions as they occur (for instance, in combustion flames). High-power lasers can create plasmas under unprecedented conditions and may provide a method for accelerating particles without the need for gigantic accelerators. The influence of lasers on science is too broad to summarize; perhaps it is sufficient to point out that lasers are now ubiquitous in laboratories of physics, materials science, chemistry, biology, physiology, and many of the other sciences.

One of the most dramatic developments in condensed-matter physics is the opportunity to carry out spectroscopy from the optical to the x-ray region by using radiation of unprecedented intensity, many orders of magnitude brighter than was previously possible. The radiation is provided by a gift from particle physics: synchrotron light sources. The fundamental technology of synchrotron sources is the technology of electron accelerators. At the same time, this gift has been reciprocated in that the high-field superconductors created by materials scientists in the 1960s have made proton accelerators in the trillion-electron-volt (TeV) range practical. Superconductor technology is also having a profound influence on health care, for superconducting magnets are an essential element of the magnetic resonance imaging (MRI) technique.

Countless other examples could be cited of the unifying ideas and techniques that link even the most disparate subjects in physics and, indeed, link physics to the other sciences and to the central technical and industrial needs of society. Appreciation of the unity of physics is essential in planning research and developing science policy.

PROGRESS IN PARTICLE PHYSICS

Quarks and Leptons as Elementary Particles

The longing to discover the most elementary particles in nature is deeply rooted in physics. At the beginning of this century, physicists discovered that the atom is not a single particle but that it consists of electrons moving rapidly around a central nucleus; in the 1930s, it was discovered that the nucleus is not a single particle but that it consists of protons and neutrons tightly bound together. Initially, the protons and neutrons were assumed to be elementary, but during the 1950s and early 1960s a large number of similar particles, called hadrons, were discovered. Over 100 hadrons are now known. In the 1960s, it was suggested that the properties of all the hadrons could be explained by recognizing that they are not elementary particles but are composed of smaller particles, each with an electric charge of one third or two thirds that of the electron. These smaller particles are called quarks.

During the past decade, the quark model has been experimentally verified. For example, jets of hadrons discovered in high-energy experiments have been explained in great detail by viewing the collision not as a collision of hadrons, but as a collision of the constituent quarks.

In the earliest version of the quark theory, there were only three different quarks—up, down, and strange. However, the discovery of the J/ψ particle in 1974 and of the Y particle in 1977 led to the addition of two new quarks—the charm and bottom quarks. Definitive evidence for a sixth quark, the top quark, is now being sought (Figure 2.2). With these six quarks, the existence of *all* the hundred-plus hadrons could be explained.

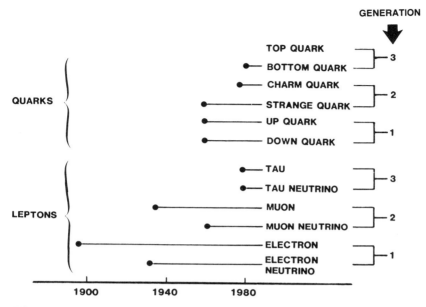

FIGURE 2.2 Quarks and leptons are the basic particles of matter. Most of them were discovered in the past two decades. There is not yet definitive evidence for the existence of the top quark.

A startling feature of quarks is that, as far as we know, there is no possibility of isolating one of them. Quantum chromodynamics, a theory of the strong interactions between quarks, accounts for this by predicting that the energy to separate two quarks grows continuously as they are separated owing to the creation of a gluon string between them. Quantum chromodynamics is now so far advanced that theorists can apply it to calculate numerically the masses of the hadrons. The gluon-mediated interaction between quarks becomes weaker the closer the quarks are together. This effect, called asymptotic freedom, makes it possible to view hadron collisions as a series of collisions between individual quarks. In addition, it eliminates some of the internal inconsistencies that plagued previous theories.

The family of quarks was discovered relatively recently. There is a second family of elementary particles, the leptons, some of whose members have been known for many decades. The first of these, the electron, was discovered at the turn of the century; two others—the neutrino and the muon—were predicted in the 1930s and discovered experimentally in the following two decades. The electron and muon are charged; the neutrino is electrically neutral. In 1963, it was found that there are at least two kinds of neutrino, one associated with the electron, the other with the muon. In 1975, a third, very heavy, charged lepton, the τ, was discovered. Soon thereafter, evidence was found for a third type of neutrino associated with the τ. Thus, we now know of six leptons, which form three groups.

A decade of experimental and theoretical research on the quark and lepton families has led to the realization that many of their properties can be explained by two simple ideas. First, the particles can all be classified into pairs by their properties and interactions. Each charged lepton is paired to a unique uncharged neutrino. Among the quarks, the up and down quarks pair together, as do the charm and strange quarks. Second, the quark family and lepton family are related: each quark pair is uniquely related to one lepton pair by a simple arrangement called the generation model.

We do not understand why the generation model works, nor do we know if there are more generations. In fact, we do not know why the quarks and leptons are related at all. To find the answers to these questions, experiments with higher-energy accelerators are being planned to explore the internal structure and dynamics of the known particles, to search for new particles, and to provide the data essential to constructing new theories.

Unification of the Forces of Nature

There are four fundamental forces in nature. Two have been known for centuries: the force of gravity and the electromagnetic force. In the period between the World Wars, two other forces were identified: the strong force, which holds the nucleus together, and the weak force, which is responsible for many types of radioactivity. Since the days of Einstein, it has been the dream of physics to develop a unified theory, a theory that describes *all* these forces with a single set of equations and concepts.

In the last two decades, the dream has been partially realized. The electromagnetic and weak forces have been combined in a single theory: the photon (a particle of light) carries or mediates the electromagnetic interactions, whereas the weak forces are mediated by massive charged particles called the W^+ and W^- and by a neutral particle called the Z^0. Interactions mediated by the neutral particle, called neutral currents, have been discovered in experiments with high-energy neutrino beams and in studies of how electrons and positrons annihilate each other. In addition, atomic physicists have detected minute effects due to neutral currents in the spectra of cesium, bismuth, and thallium, adding to evidence for the theory. In one of the

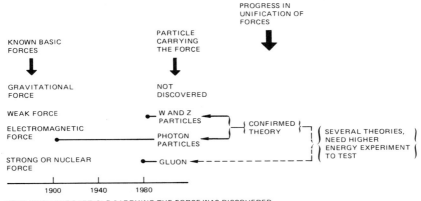

FIGURE 2.3 The four known basic forces expected to be carried or mediated by an elementary particle. The particles carrying two of the forces have been discovered in the past decade. Since the days of Einstein, physicists have wanted to unify the forces so that they can all be derived from a single basic equation. That has been accomplished for the weak and electromagnetic forces.

most ambitious experiments ever undertaken, the W^+, W^-, and Z^0 particles have recently been observed, eliminating any further doubts about the origins of the weak force (Figure 2.3).

Efforts to unify the electroweak forces and the strong force are leading to exciting new challenges in theoretical and experimental physics. Theories have been proposed predicting that the proton is not stable but will decay and that magnetic monopoles exist. Testing these predictions is one of the many opportunities for particles physics in the coming decade.

PROGRESS IN NUCLEAR PHYSICS

The challenge to understand the diverse arrangements of protons and neutrons in the nuclei of atoms has fascinated physicists since the 1930s. In the 1960s, the model of the nucleus as a simple collection of protons and neutrons evolved into a more complex picture in which the strong nucleon-nucleon interactions arose from the exchange of mesons; now this picture is being replaced by the rich portrait that emerges from recognition of the underlying quark-gluon nature of nucleons. One day it should be possible to explain the entire nucleus as a many-body system of interacting quarks and gluons. The experimental and theoretical challenge is enormous, but so is the reward of understanding nuclear matter (Figure 2.4).

FIGURE 2.4 Streamer chamber photograph showing catastrophic nuclear collision between an argon-40 (heavy-ion) projectile with 1.8 GeV/nucleon energy (total kinetic energy, 72 GeV) and a lead target nucleus. Such events are photographed and analyzed to extract information on extreme conditions of temperature and density achieved in these collisions. The charged particles that are produced are deflected by a magnetic field that surrounds the 1.2-m-long chamber. New ultrarelativistic heavy-ion accelerators would use similar techniques to explore a new phase of matter—the quark-gluon plasma. (Courtesy of the Lawrence Berkeley Laboratory and GSI Darmstadt.)

The first steps toward this goal have already been taken. Meticulous studies of the deuterium nucleus (a proton plus a neutron) reveal that, when the two nucleons are close together, they may be best described in terms of their six constituent quarks. Similar studies in iron (56 nucleons) reveal that the distribution of quarks in iron may be different from their distribution in deuterium. The quarks in iron seem to be able to move among the individual nucleons; apparently, in a large nucleus, quarks are less strongly confined than they are in a small nucleus.

In the 1940s, it was discovered that nuclei could vibrate. In the simplest mode, the protons and neutrons move in opposite directions. Other types of vibration have been predicted theoretically, and, within the past 10 years, they have been discovered experimentally. Observation of the breathing mode represents a major advance in understanding the basic properties of nuclear matter, because the measurements of the breathing mode have made it possible to determine the compressibility of nuclear matter.

Many other modes of motion have been observed. They range from collective vibrations (such as the breathing mode) to modes in which a single nucleon is excited from one energy level to another. Between these extremes, resonant states have been found in which the nucleus behaves as if it were composed of two separate smaller nuclei making up a nuclear molecule and in which the nucleus rotates so rapidly that it is close to flying apart.

The most successful model of the nucleus has been the shell model, in which nucleons fill orbits much as electrons do in atoms. With electron-scattering experiments, it is now possible to observe individual shell-model orbits. A dramatic technique for probing nuclei is to transform a neutron suddenly into another particle, one that the Pauli exclusion principle cannot exclude from the many orbits that are already filled. A K^- meson, for instance, can be used to convert a neutron inside the nucleus into a hyperon, such as a Λ or Σ^0 particle. Special facilities have been developed for the systematic study of hypernuclei. By measuring the energy-level structure and the gamma-ray decays of hypernuclei, it has become possible to study the forces that bind the hyperons in the nuclei.

Experimental nuclear research is advancing toward the study of nuclei in states of higher excitation energy and higher angular momentum and toward more exotic (neutron-rich or proton-rich) nuclei farther from the valley of stability. Near the limits of nuclear stability, the balance between attractive and destructive forces is so delicate that the nuclei provide sensitive testing grounds for theories of nuclear structure. New nuclei at the limits of stability have been discovered. In 1974, element 106 was created and identified in an experiment where only one atom was produced in 10 billion collisions. Recently, elements 107 to 109 were discovered.

A major effort of nuclear research is the study of collisions of nuclei at very high energies. The collisions may result in the creation of a quark-gluon plasma. Such a discovery would be of extraordinary importance for quantum chromodynamics, and it could have significant implications for astrophysics and cosmology as well.

Precise calculations of the properties of small nuclei have made it possible to test the effects of meson exchange currents and to recognize the quark-gluon nature of nucleons. New insights into nuclear structure have been gained recently from a theoretical model in which the nucleons are paired to make bosons (particles such as the deuteron with integral spin), which are then used as the building blocks to describe the energy levels and how they decay. This model, the interacting boson model, has succeeded in correlating vast amounts of data and has proven to be helpful in suggesting new studies. There is hope that new symmetries, called supersymmetries, can be found by coupling fermions (particles with half-integral spin) to this boson model.

PROGRESS IN CONDENSED-MATTER PHYSICS

During the 1950s and 1960s, physicists explored the electronic properties of crystalline solids and constructed a comprehensive picture of electron energy levels, transport mechanisms, and optical properties of most simple metals, insulators, and semiconductors. Today, condensed-matter physics concentrates primarily on surfaces and interfaces, systems with strong fluctuations (including turbulence), and systems with varying degrees of disorder.

Surfaces, Interfaces, and Artificially Structured Materials

Our understanding of surfaces has evolved differently from our understanding of crystalline solids. Photoemission studies (that is, the analysis of light-induced emission of electrons) have provided extensive information on the electronic energy spectra of surfaces. However, our knowledge of the atomic structure is still relatively meager. Because the freedom of motion on a surface is large, surfaces often reconstruct to satisfy bonding restrictions in complex ways, resulting in a rich variety of surface structures. New techniques are starting to reveal the secrets of surface structures and their phase transitions. These techniques include the tunneling microscope, which makes possible pictures of surfaces with angstrom (0.1 nanometer) resolution, surface electron microscopy, surface x-ray scattering using synchrotron radiation, and various atom- and ion-scattering methods. In addition, the traditional method of low-energy electron diffraction has been greatly refined. These advances would not have been possible without the continual improvement in ultrahigh-vacuum technology and the use of computer control.

Theoretical understanding of surfaces has expanded rapidly. New numerical techniques have been developed to take advantage of the incredible computational power provided by modern computers. Electronic structures can now be calculated for different positions of the atom, and their relative stabilities can be examined. In this way the equilibrium atomic arrangement can be determined, at least for simple surfaces. An interesting feature of these calculations is that a relatively simple approximation to the correlations between electrons—the local density approximation—seems to give excellent values for the total surface energies. This approximation is being used to predict bulk phases and detect configurations as well.

As the technology developed for preparing clean, well-characterized surfaces, it became possible to control atom-by-atom deposition on a surface. By gradually laying down planes of one type of atom followed by planes of another type, one can create a new class of materials—artificially structured materials. The classic example of such a material is one in which layers of gallium arsenide are alternated with layers of gallium aluminum arsenide. Because the layers differ markedly in their electronic properties, the multilayer materials can exhibit unusual electronic behavior. For example, mobile electrons can be confined in one type of layer so that they move only in two dimensions. New electronic devices, including semiconductor lasers, are being made using these materials (Figure 2.5).

Studies of electrons confined to move in only two dimensions have led to the discovery of the quantized Hall effect. In this effect, the Hall conductivity that is associated with the current flowing perpendicular to both the magnetic field and the electric field is quantized in units of the square of the electron charge divided by Planck's constant. This relation appears to hold accurately irrespective of the material used. This totally unexpected discovery was followed by the discovery of the fractional quantization of the Hall current, whose existence has been attributed to a completely new correlated state of matter.

Phase Transitions and Disordered Systems

Until recently, there was no fundamental understanding of the properties of a material as it undergoes a phase change, such as the

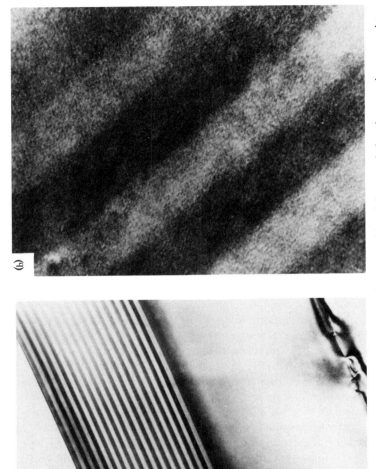

FIGURE 2.5 Electron photomicrographs of artificially structured materials. (a) Low-resolution electron microscope picture of artificially structured material consisting of alternating layers of $GaAs_{0.2}P_{0.8}$/GaP that are 2.28 nm thick. These layers were grown on a $GaAs_{0.1}P_{0.9}$ substrate by a technique called metalorganic chemical vapor deposition (MOCVD). (b) High-resolution electron microscope picture of atomic rows in artificially structured material of alternate layers of $GaAs_{0.15}P_{0.85}$/GaP similar to (a). Layers are 1.2 nm thick. (Grown by R. M. Biefield, electron microscope pictures by C. Hills, both at Sandia National Laboratories.)

abrupt vanishing of spontaneous magnetism of a ferromagnet at high temperatures. In the early 1970s, careful experimental studies, combined with the theoretical development of scaling laws and a technique called the renormalization group approach, led to the creation of an accurate procedure for calculating the properties of materials near a phase transition. The most striking feature of systems near a phase change is the strong fluctuation in their properties as the new phase builds up. These fluctuations, and the singular behavior at a phase transition, are now understood in detail, and a wide variety of phase changes has been analyzed.

Because the renormalization group and scaling ideas accurately describe fluctuations, they constitute versatile tools for understanding other phenomena that involve strong fluctuations. Perhaps the best example of a completely different phenomenon to which these techniques have been applied is localization. When electrons move in a potential that varies randomly in space, they can be localized in a well of the potential, provided that the variations of the potential are sufficiently large. Otherwise, the electrons will be free to move, as in a conductor. We now have a detailed theory for how the transition between these two regimes takes place and under what circumstances localization can occur.

Disordered materials and glasses are attracting wide interest. Among the new discoveries in glasses is a high density of low-energy states called tunneling states. The existence of such states is well established, but their microscopic origin is not understood. Similarly, the criteria for metastability of glassy materials are not well understood. Another area attracting attention is the study of partially ordered systems, such as liquid crystals in which the molecules maintain some degree of positional or orientational order but are not bound to specific sites. The liquid-crystal displays that are seen everywhere are one of the technological innovations from research on their fascinating properties.

Physicists have studied a system that is somewhat simpler than real glasses called the spin glass. Discovered in the early 1970s, a spin glass is a state of matter in which the magnetic spins of randomly located atoms freeze in direction at low temperatures. These systems appear to be in many ways analogous to real glasses. Spin glasses and related systems introduce the new feature that, below a well-defined temperature, the properties are forever history dependent.

The research on spin glasses has contributed to developments in several other fields. One is a spin-glass model of the neural networks in which the spin directions are analogous to *on* or *off* states of neurons, and different spin configurations represent different memories. Another is the development of the Monte Carlo annealing techniques, derived from numerical simulations of spin glasses. This technique is being applied to such problems as the most efficient way to wire circuits and the determination of molecular configurations. The studies of partially ordered systems will help to advance many areas of science and technology.

PROGRESS IN ATOMIC, MOLECULAR, AND OPTICAL PHYSICS

The physics of atoms, molecules, and light underlies our understanding of the world about us. Research in this area has advanced rapidly during the past decade, propelled by a host of new techniques based on lasers and nonlinear optics, by other experimental methods such as supersonic molecular beams, particle traps, clusters, and highly charged ions, and by new theoretical concepts and calculational techniques.

Laser spectroscopy of atoms and molecular systems has achieved unprecedented resolution higher than traditional means by a factor of one million. However, the term laser spectroscopy has also come to signify a much wider area of research: it encompasses the creation and study of species such as free radicals and molecular ions; the development of nonlinear techniques such as coherent anti-Stokes Raman scattering (CARS) that make it possible to monitor chemical reactions as they take place in a combustion chamber; and the development of new metrological methods (for instance, the optical frequency-counting technique that recently led to the redefinition of the meter as the distance traveled by light in a specified time interval). Lasers make it possible to watch how energy is transferred in molecular collisions. With femtosecond lasers, it will be possible to take a "snapshot" of a molecule as it undergoes a reaction and to observe how molecular vibrations decay on surfaces. Developments in short-wavelength lasers and nonlinear optics are opening the way to the creation of intense laboratory sources of far-ultraviolet radiation and soft x rays.

Electromagnetic traps have been designed that can store electrons, positrons, or ions up to months at a time, providing a new arena for precision spectroscopy and for studying collisions. With such a trap, in an experiment that used only a single particle at a time, the magnetic moment of the electron has been measured to an accuracy of 40 parts per billion, a milestone in precision measurement. In conjunction with the results of major theoretical and calculational efforts, the research provides one of the most exacting tests of the theory of quantum electrodynamics. Ion traps have made it possible for the first time to study reactive collisions between cold ions and molecules; such collisions are important to our understanding of chemical processes in interstellar space (Figure 2.6). These traps are also being employed in the study of collective motion in a charged plasma, in new types of atomic clocks and optical frequency standards, and in sensitive tests of the isotropy of space.

The Lamb shift of hydrogen—the shift in energy levels due to intrinsic fluctuations in the electromagnetic field—has been measured to such high precision that the comparison between experiment and theory is now limited only by our understanding of the internal structure of the proton. Leptonic atoms, short-lived hydrogenlike atoms in which the proton is replaced by a positron (positronium) or muon (muonium), are not affected by proton structure. Positronium has been studied by laser spectroscopy, and the Lamb shift in muonium has recently been observed. A new arena for the study of quantum electrodynamics has been opened by the development of techniques for laser and precision x-ray spectroscopy of highly charged hydrogenlike ions.

The spectrum of the most elementary negative ion, H^-, has been studied by directing laser light against a relativistic beam of the ions. Owing to the ions' high speed, the color of the laser light was changed by the Doppler effect from green to ultraviolet. The experiment demonstrated the existence of an electronic resonance structure that had been predicted theoretically. This is one of the central problems in atomic physics; other new techniques that have been brought to bear on it include the theory of collective coordinates, high-resolution electron scattering, photoionization with synchrotron and laser light, high-energy ion scattering, and multiphoton spectroscopy.

In high-energy ion-atom collisions, vacancies in the innermost electron shells have been discovered to be created by the promotion of electrons in the quasi-molecule that is formed during the collision. X rays from transitions between orbitals of the transient molecule have been seen. The results are closely analogous to molecular behavior previously observed in outer electrons during low-energy collisions. These x rays are important in the deposition of energy within biological material by heavy ions and in ion-beam compression of fusion pellets.

During heavy-ion collisions, an enormous electric field is produced

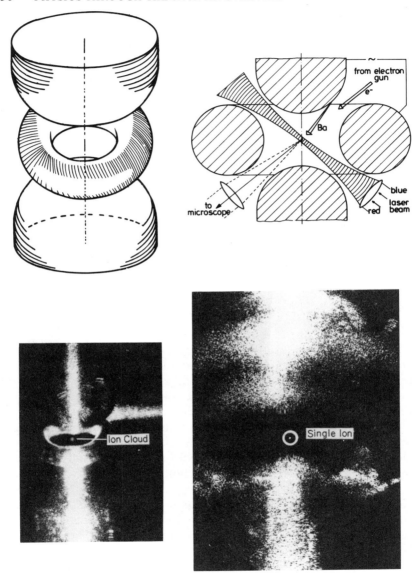

FIGURE 2.6 Ions can be trapped in high vacuum, using static and oscillating electric fields, and viewed by laser light. The experiments can be so sensitive that single ions can be observed under close to ideal conditions of isolation. In this experiment, barium ions are formed in the center of the doughnut-shaped electrode by bombarding barium vapor with electrons. The ions are observed by their fluorescence under laser light. The photograph at bottom left shows the laser light scattered by a small cloud of trapped ions.

in the vicinity of the superheavy nucleus that briefly exists; it can be so large that the binding energy of an inner-shell electron exceeds a million electron volts (1 MeV), twice the electron's rest mass. In such a field, an electron-positron pair can occur spontaneously, a process that is now believed to have been detected.

PROGRESS IN PLASMA AND FLUID PHYSICS

Progress in Plasma Physics

Plasmas constitute a state of matter in which most of the atoms are broken down into free nuclei and electrons. Interest in plasmas arises from their unique physical properties, the variety of roles they play throughout the universe, and their many applications. Plasmas amplify electromagnetic waves by collective nonlinear effects and display turbulent behavior. Electromagnetic coupling can confine a plasma for a possible fusion reactor, using the same fundamental process that governs the formation of sunspots and the structure of planetary magnetospheres.

Because plasma behavior is inherently nonlinear, it is difficult to calculate by conventional analytic techniques. Consequently, numerical analysis and modeling are widely employed. Simulations containing hundreds of thousands of particles have been used in one of the largest and most ambitious endeavors of simulation physics. The National Magnetic Fusion Energy Computer Center, linked to the major fusion centers, has been established to pursue this research.

Mastery of plasma physics at the level needed to understand fusion and space plasmas requires a complete synthesis of classical electrodynamics and nonequilibrium statistical mechanics. Plasma research has led to a resurgence of interest in classical physics, and it has stimulated a great deal of activity in applied mathematics.

Large-amplitude space-charge waves are one example of the many

---

In the blown-up photograph at bottom right, the light scattered by *one* barium ion can be discerned in the circled region. Laser light can also be used to cool the ions, reducing the energy-level shifts due to the second-order Doppler effect. Trapped-ion methods are being applied to ultrahigh-resolution optical spectroscopy and to the creation of new types of atomic clocks. The methods are employed to study collisions and chemical reactions, including reactions at very low temperature, and to study collective motion in charged plasmas. (Courtesy of the University of Hamburg, Federal Republic of Germany.)

varieties of nonlinear phenomena displayed by plasmas. Such waves can generate electric fields up to hundreds of millions of volts per centimeter. They can be used to trap and accelerate particles, and ions have been observed to acquire energies as high as 45 MeV. The research is being pursued in the hope of accelerating ions to energies as high as one billion electron volts (1 GeV) in extremely short distances. Solitons—large-amplitude waves that hold together against normal dispersion—provide another example of nonlinear plasma phenomena.

A notable advance in plasma physics is the solution to the problem of magnetic-field reconnection. This is the mechanism by which magnetic field lines reconnect on either side of a current sheet as the field dissipates. This understanding is important not only for laboratory plasmas but for space plasmas such as those involved in solar magnetic activity and the Earth's magnetosphere. Another significant advance is the discovery that direct currents can be driven by applying radio-frequency fields to plasmas in a tokamak device. The discovery may allow a tokamak confinement configuration to operate in steady state, rather than in a pulsed mode, an achievement of potentially enormous economic importance.

Studies of relativistic electron beams moving in oscillating transverse magnetic fields have led to the development of new coherent sources of radiation such as the relativistic magnetron, the gyrotron, and the free-electron laser. These devices produce radiation from microwave frequencies through the infrared to the visible at power levels that can be extraordinarily large. They are expected to find applications in science, industry, defense, and medicine.

Fusion

Magnetic fusion research has made rapid technical progress during the past decade. One approach—toroidal magnetic confinement as occurs in tokamak reactors—today stands at the threshold of satisfying the requirement for energy breakeven in deuterium-tritium plasmas. The progress was possible because of breakthroughs in understanding the nature of collective modes in plasmas and advances in plasma control and heating. An important result of the work was the empirical discovery that energy confinement in ohmically heated plasmas is proportional to the plasma density and volume. Along with the discovery that radio-frequency waves can drive direct currents, this advance marks a significant step toward the eventual economic success of a tokamak reactor.

Alternatives to toroidal confinement systems are also being developed. One approach is the mirror configuration, which employs a linear magnetic-field geometry that pinches the ends to form mirrors for the plasma particles. The tandem mirror concept, in which the electric fields are generated along magnetic field lines, has been introduced to suppress plasma leakage through the ends. Results point toward the possibility of mirror confinement systems adequate for fusion-reactor applications. Multimegawatt neutral-beam sources have been developed to fuel the mirror machine and heat plasmas to fusion temperatures. Two-hundred-million-degree plasmas at fusion-plasma densities have been achieved. In toroidal devices, 80-million-degree temperatures have been obtained (Figure 2.7).

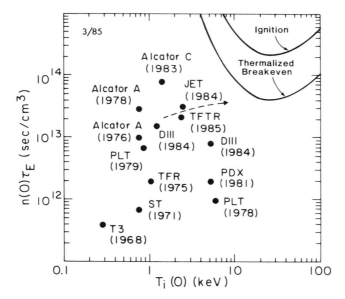

FIGURE 2.7 Plot of Lawson confinement parameter $n(0)\tau_E$ versus central ion temperature $T_i(0)$ for several tokamaks. Here, $n(0)$ is the central density and τ_E is the energy confinement time. The year of the result is indicated in parentheses. The JET (Joint European Torus) and TFTR (Tokamak Fusion Test Reactor) tokamaks, with auxiliary heating, are expected to operate in the 10- to 15-keV range during 1985-1987. The thermalized breakeven and ignition curves refer to an equidensity fuel mixture of deuterium and tritium plasma with Maxwellian ions.

Space Plasmas

The interactions of the Sun's wind and its magnetic field with the magnetic fields of the planets are but one example of the plasma phenomena that occur throughout our solar system. The magnetic field and plasma surrounding each planet define a region known as the magnetosphere. The Earth has a relatively quiet magnetosphere, which we are understanding in increasing detail. For example, we are beginning to understand how the reconnection of the Earth's magnetic field lines is related to auroral activity. Data from the Pioneer and Voyager space missions have yielded a detailed picture of planetary magnetospheres and the electromagnetic activity in the solar system (Figure 2.8). Today, space probes are providing data on collective oscillations, shocks, particle acceleration, and instabilities. Outside the solar system, plasma behavior in extreme astrophysical environments can give rise to such bizarre phenomena as the jets of particles in opposite directions that have been observed to be ejected from pulsars.

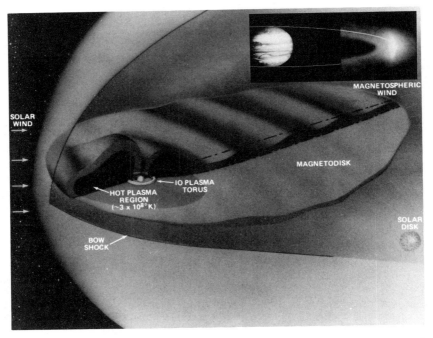

FIGURE 2.8 The magnetosphere of Jupiter. The insert shows a Voyager photograph of Io's sulfur plasma torus. (Courtesy of NASA.)

Fluid Physics

The physics of fluids is far from understood because fluid motion involves many degrees of freedom and is inherently nonlinear. Understanding fluid flow, whether the fluid is a gas or a liquid, is essential for applications such as weather prediction, flight and transportation, plate tectonics, combustion and chemical reactions in flames, and biological problems such as blood flow in cardiovascular systems. Thus, any advance in our understanding of flows, particularly turbulent and unsteady fluid flows, can be expected to have enormous technological impact. For example, recent advances in the theory of acoustic damping and turbulent flow, applied to jet noise, led to a thousandfold reduction in the acoustic energy emitted by aircraft, providing a major reduction in perceived noise levels.

There is increasing interest in the so-called nonideal fluids. New constitutive models for fluids, based on molecular structure, have led to a better understanding of the striking flow properties of polymer solutions and drag-reducing agents. These fluids have an interesting application in fire-fighting equipment; the addition of minute quantities of very long macromolecules greatly increases water flow and reduces steam backup.

Progress toward understanding the onset of turbulent flow is encouraging. Developments include an increased understanding of chaotic behavior in simple systems, new methods for observing fluid behavior near onset, and new techniques for analyzing the data. The advances are due in large part to modern large-scale computational capabilities. Among the potential applications of the work is the prediction of global-scale flows for both short- and long-term weather forecasting.

PROGRESS IN GRAVITATION, COSMOLOGY, AND COSMIC-RAY PHYSICS

Gravitational Physics

The best-known prediction of Einstein's General Theory of Relativity is that gravity bends light, but the theory predicts other equally startling effects. They are all so small, however, or so hard to observe, that testing general relativity presents a formidable challenge. During the past decade, there has been a breakthrough due to space techniques. For example, the propagation time for light passing the Sun has been monitored using a spacecraft. General relativity predicts that light is not only bent by the Sun's gravity but also slowed by it; a beam grazing the Sun is delayed by 250 microseconds. By using the Viking Lander spacecraft on Mars, this tiny delay has been measured to an accuracy of 0.1 percent.

Space techniques have provided other tests of gravitational theory. A hydrogen-maser atomic clock in a rocket was compared with a similar maser on the ground, allowing the minute effect of the Earth's gravity on clock rate (the gravitational redshift) to be measured to an accuracy greater than 1 part in 10,000. The results agreed with the value predicted by the equivalence principle, which relates the effects of acceleration and gravity. Cosmological arguments have opened the possibility that Newton's gravitational constant G may not really be a constant but that it changes as the universe ages. Ranging measurements using the Viking Lander on Mars have been combined with other solar-system data to set a limit on the possible change: it is no more than 1 part in 10^{11} per year.

One of the most dramatic quests in gravitational physics today is the search for gravitational radiation. The radiation from known sources is predicted to be so weak that detecting it requires highly innovative experimental techniques. One method attempts to sense the passing of gravitational waves by their effect on the length of a large aluminum bar that is cooled to liquid-helium temperature and carefully isolated from vibration. A strain level (fractional change in length) of 10^{-18} can be detected. The extreme sensitivity of this measurement can be appreciated by noting that a strain of 10^{-18} in a 1-meter-long bar is a change of length by 0.1 percent of the diameter of an atomic nucleus. Further improvements are under way. Another approach uses laser interferometers whose mirrors are mounted on inertial platforms. These detectors are expected to reach strain levels of 10^{-23} when baselines of several kilometers are achieved. Interest in gravitational radiation goes beyond its role in gravitational theory; the waves can reveal sources, like black-hole formation, that are invisible to us now. The discovery of gravitational radiation would truly open a new window on our universe.

A compelling demonstration of the reality of gravitational radiation has been provided by careful observations of a system of two compact objects, one of which is a pulsar that emits regular pulsed signals. The 8-hour orbit has been studied in exquisite detail by clocking the radio pulses. Since 1975, the orbit has decayed owing to the loss of energy by gravitational radiation; theory and observation agree to within 1 percent.

Fundamental advances in relativity theory have accompanied these experimental advances. The positive energy theorem has proved that, in general relativity, any isolated system must have positive total energy. This is by no means obvious, because gravitational binding energy is negative. In another milestone discovery, theoretical relativists have shown that black holes evaporate by emitting thermal radiation, the temperature being inversely proportional to the mass. They have also shown that black holes have well-defined entropy and that a generalized form of the Second Law of Thermodynamics is valid even for systems containing black holes.

Cosmology

Confidence in the theory of the primordial explosion—the big bang—continues to increase. New measurements of the spectrum of the 3-K radiation that fills the universe, as well as recent measurements of the cosmic abundances of the light elements, match the big-bang predictions. Equally important, whatever direction one looks in, the 3-K radiation is found to be remarkably uniform: the universe is apparently isotropic to better than 0.01 percent. This isotropy confirms Einstein's assumption of cosmic homogeneity (the cosmological principle), but it presents a puzzle in causality. The regions of space being viewed had not yet been connected by light signals at the time of emission; so lacking any possibility of communication, how could these regions "know" the temperature elsewhere?

The extreme temperatures predicted for the earliest moments of the big bang correspond to energies far beyond the wildest possibilities for elementary-particle accelerators, but they can nevertheless be conceived in the imagination of theorists. If we calculate the cosmological consequences of theories like the Grand Unification Theory (Figure 2.9) and compare them with observations, both particle physics and cosmology are advanced. Possible explanations are being found not only for the large-scale causality puzzle but for problems such as why the ratio of baryons to photons in the universe, only 10^{-9}, is so small.

Theoretical cosmology is advancing rapidly on many fronts. Important progress has been made toward understanding the formation and evolution of large-scale structure—galaxies and clusters of galaxies, for example. Such studies may lead to an estimate of mass distribution early in the life of the universe. Theoretical ideas for possible dark-

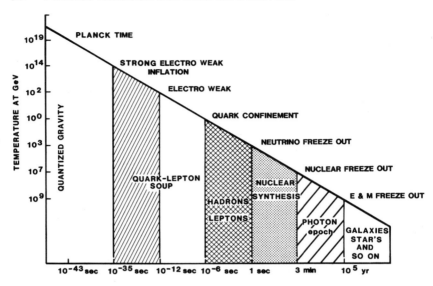

FIGURE 2.9 Schematic illustration of current theory of events involved in the big-bang theory of the universe. The abscissa is time, and the ordinate is temperature (measured in GeV, where 1 GeV is approximately 10^{13} K). We cannot say much about the earliest times because we do not understand physics at such high energies. As the temperature decreases, the various forces separate into the forces as we know them today. (1) At 10^{-35} s the strong force separates from the electroweak forces. (2) At 10^{-12} s the electromagnetic and weak forces separate. (3) At 10^{-6} s the quarks become confined into hadrons. (4) At 1 s the thermal neutrinos separate from the other particles in the universe. (5) At 3 minutes the nuclei freeze out. Finally (6), at 10^5 years the photons making up the 2.7-K blackbody radiation of the present universe cease to interact with matter. Subsequently, the galaxies and stars are formed.

matter components run from "ordinary" astrophysical objects—very-low-mass stars or black holes—to exotic new particles—axions, massive neutrinos, and photinos. Whatever the answer, it will have profound implications for our understanding of the past, present, and future of the universe.

Cosmic-Ray Physics

As in cosmology, scientific spacecraft have opened many new possibilities in the study of cosmic rays. The ability to measure abundances of individual isotopes of the lighter elements, such as neon, magnesium, and silicon, and elemental abundances of rare extremely heavy elements has indicated that galactic cosmic rays may well be accelerated interstellar material. Their composition is similar to, but significantly different in detail from, either that found in the solar system or that resulting directly from nucleosynthesis in supernovae. Information from deep-space probes has increased our understanding of the interaction of galactic cosmic rays with the local interplanetary environment.

The highest-energy cosmic rays are observed using ground-based and underground detectors. Cosmic rays with energies of as much as a joule have been observed, and the isotropy in the arrival direction of these highest energy particles has indicated that they have an extragalactic origin. In addition, discrete sources of gamma rays with energies up to 10^{15} eV have been observed, and they indicate a few localized cosmic-ray accelerators with enormous power. Understanding the acceleration mechanisms for such energetic particles is a major goal of current research.

INTERFACES AND APPLICATIONS

Physics is woven so tightly into the fabric of science and society that the boundaries separating it from neighboring sciences are hardly discernible. In the wider context of society as a whole, hardly any aspect of modern life remains unaffected by the discoveries of physics.

Interface Activities

The basic concepts of physics are incorporated in virtually every area of science. In addition, physics interacts directly with many sciences by exchanging theoretical approaches and experimental techniques.

CHEMISTRY

The interface between physics and chemistry is among the best developed interdisciplinary areas in science. Advances in spectroscopy, including laser spectroscopy, nonlinear optics, extended x-ray absorption fine structure (EXAFS), the advent of synchrotron light sources, and molecular beams are playing increasing roles in chemistry. These advances have already had a significant influence on molecular physics and surface chemistry. They are also finding applications in other areas of growing physical-chemical interest (such as the study of polymers and liquids) and in photochemistry and photochemical processing. Order-disorder transitions—a central problem in physics—are being studied in the chemical arena using media such as micellar and colloidal crystals, microemulsions, and liquid crystals. Condensed-matter physics is benefiting from the new classes of materials with unique physical properties that are being created by organic chemistry. For example, the discovery of organic conductors of electricity in the early 1970s has led to the creation of effectively one-dimensional conductors that display the drastic fluctuation effects predicted by one-dimensional statistical mechanics. By manipulating the molecular architecture, one can pass systematically from insulators to semiconductors to metals. We can look forward to numerous applications of these discoveries, particularly in materials science and electronics.

BIOPHYSICS

Physics encompasses increasingly complex problems as its experimental and theoretical powers expand. In biophysics, for example, new frontiers have been opened by the development of methods to observe the conductance of single molecular channels through biological membranes. Hundreds of different channels have been identified, with more surely to be discovered. Each channel is switched in a quasi-random process while it is simultaneously modulated by mechanisms such as binding of neurotransmitter molecules. One can look forward to working out the finest details at the molecular level in the next few years. Molecular genetic technology will assist this research by creating mutants that are specially engineered to probe the molecular mechanisms. The secrets of protein structure and enzymatic processes can be expected to unfold as they are probed with synchrotron radiation, x-ray and electron-beam crystallography, laser Raman spectroscopy, magnetic resonance, and other tools from physics. The new, powerful techniques of biophysics and molecular genetics promise to deepen our understanding of molecular processes in biology at all levels.

GEOPHYSICS

Experimental and theoretical methods of physics are contributing to geophysical understanding of the Earth's three phases—atmosphere, oceans, and solid planet. Today, solid-earth geophysics is dominated by the concepts of plate tectonics. Modern seismology and innovative techniques of metrology, particularly those using lasers and atomic clocks, have made it possible to monitor terrestrial motions with unprecedented precision. Extended movements of the solid-earth crust can now be interpreted with such high precision that one can begin to account for fine details like mineralization. In the atmosphere, the analysis of the turbulent fluid flow associated with global weather patterns is being revolutionized by advances in large-scale computing. The fundamental theory of turbulence, however, has yet to be created; the lack of such a theory continues to limit progress. In the ocean, underwater seismology, sound-propagation tomography, and satellite sensing are providing data on the ocean's temperature, its level, the state of the sea, and the strength of the currents, all of which are essential for basic understanding of this intricate system. The problems of geophysics present a formidable challenge because of their inherent complexity. However, it is vital that progress continue, because geophysics is involved in the utilization of every type of energy resource and is an essential element of any attack on the global energy problem.

MATERIALS SCIENCE

The systematic study of materials and the development of techniques to control, modify, and create materials are central goals of this interdisciplinary subject. To this effort, physics contributes fundamental theory, such as the rapidly developing theory of disordered materials. It also contributes new experimental tools, such as synchrotron light sources and the free-electron laser. Microelectronics would not have been possible without the techniques for making ultrapure semiconductors developed by materials scientists; a strong underpinning of physics and chemistry was essential to this advance. Conversely, the precisely controlled materials fabricated by materials scientists constitute an invaluable resource for condensed-matter physics. The development of methods for making artificially structured materials is one of the many areas in which condensed-matter physics and materials science are moving forward together. These materials provide a testing ground for basic physical theory and a likely arena for discovering novel physical behavior. At the same time, they offer the possibility of creating materials with unique properties for applications such as ultrahigh-speed microelectronics. The development of low-loss optical fibers and the creation of the fiber-optics industry are examples of the dramatic advances made possible by the confluence in materials science of basic science, applied science, and technology.

Applications

The most profound technological advances from scientific research have often been unanticipated: no one could predict that experiments on amber and lodestone would lead to the replacement of water wheels by electrical power or that research on the magnetic interactions of nuclei in matter would lead to a revolutionary technique for medical imaging. Nevertheless, much of today's scientific research in the United States is motivated by the need to solve specific problems facing the nation. Physics plays an important role in this task. The applications of physics research to societal needs are too broad even to summarize, but the following are a few examples.

ENERGY AND THE ENVIRONMENT

Society's welfare depends on its energy supply. Physics is involved in almost every aspect of research on the generation and efficient use of energy. Progress in fusion energy has been described; fission and other technologies from physics (such as photovoltaics) are expected to play increasingly significant roles in the future. For the present, combustion remains our major source of energy. New diagnostic tools and new materials can make combustion more efficient, and even a small increase in efficiency will have an enormous economic impact. Every manufacturing process and every technology for producing energy somehow affects the environment. Understanding and controlling these effects are crucial to our future well-being. To this task physics brings essential data, analytical techniques, and theoretical tools for monitoring the earth, the oceans, and the atmosphere.

MEDICINE

Physics is addressing increasingly complex problems of biophysics and physiology. In addition to basic research, however, it contributes directly to the quality of medical care by providing new analytical tools, diagnostic techniques, and therapies. X-ray tomography has had a major impact on x-ray diagnostics; magnetic resonance imaging (described in Chapter 1) is widely regarded as a revolutionary advance in medical diagnostics. Ultrasonic imaging is yet another of the noninvasive diagnostic tools from physics. Lasers are finding increasingly widespread applications in medicine. Laser surgery replaces some highly delicate or traumatic operations with simple and straightforward procedures. Fiber-optic endoscopes exemplify the many new diagnostic instruments. Combined with lasers, the endoscopes can be used to provide new therapies that may replace elaborate surgical procedures. These examples represent but a few of the many new instruments and technologies from physics that are today enhancing the quality of health care in this nation and elsewhere.

NATIONAL SECURITY

Physics is without peer as a source of discoveries that have an impact on national security strategies and tactics. The profound effects of physics are apparent in the development of weapons systems and strategic defense systems and in the complex process of arms control. Lasers, for example, are now widely used for communications, guidance, and surveillance. The free-electron laser and cyclotron-resonance maser are capable of providing intense coherent radiation from microwave through ultraviolet wavelengths, with a potentially major influence on radar technology, particularly "stealth" technology, and on optical countermeasures. The ability to respond appropriately to a rapidly changing situation can depend on the real-time ability to acquire and interpret vast amounts of data: new optical and information-processing techniques are at the heart of this effort. In these and countless other ways, the discoveries from basic physics in former years, combined with the efforts of scientists and engineers working today, are helping to assure that the nation can meet its national security goals.

INDUSTRY

Physics contributes broadly to industry through the creation of new materials, instruments, and technologies. Beyond these, discoveries in basic research can lead to the creation of industries such as microelectronics and laser optics. Microelectronics has made possible the information revolution that is transforming society. Laser optics is revolutionizing communications and printing. Laser-assisted manufacturing, particularly when applied to robotics, is spreading through every kind of industry. As described in *Global Competition*, the Report of the President's Commission on Industrial Competitiveness (U.S. Government Printing Office, Washington, D.C., January 1985), ". . . basic research in the Nation today is a critical factor in our long-term preeminence."

3

Maintaining Excellence

Our nation excels in physics. Since World War II, the United States has played a leadership role in essentially every area of physics, and our research has won the admiration of scientists everywhere. Excellence in physics, however, is fragile. It requires a fortunate combination of circumstances: a talented and well-educated population of scientists, a society that is interested in and appreciative of new discoveries, institutional structures that give scientists the freedom to follow wherever science leads, open lines of communication among fellow scientists everywhere, and the economic resources for carrying out research at the frontiers of knowledge. Excellence in physics also requires harmony between the aims of science and the goals of society.

We cannot take for granted the continued excellence of physics in the United States. Many of the same social and economic pressures that have affected our nation have also affected physics. Career patterns and professional opportunities for physicists have changed; reduction of support for basic physics in favor of mission-oriented research has caused problems in some areas; and the need for large facilities is generating increased economic pressures.

This chapter addresses the problem of maintaining the quality of physics in the United States. Underlying the discussion is the assumption that continued excellence is essential to our national interests. To recapitulate the arguments presented in Chapter 1, physics is vital to the nation for the following reasons:

- The longing to understand nature and the cosmos is deeply rooted in mankind, and, in our time, physics has taken a profound step toward realizing this longing. By making our appreciation of nature and life stronger, the discoveries of physics enrich all society. Our achievements in science help the United States to maintain its role as a world leader because the achievements are respected by nations everywhere.
- Physics is a central discipline. The concepts of physics, and the techniques and instruments developed in its laboratories, have been widely adopted by the other physical sciences, by the life sciences, and by medicine. Excellence in physics in this nation contributes broadly to the quality of science and medicine throughout the world.
- The nation requires technologies of an ever-higher level because without them our economy cannot flourish. The world demands new technologies to sustain and enhance the quality of life in the face of increasing population and the depletion of natural resources. We must be prepared to create the technologies needed at home and abroad. Basic science is the driving force behind new technology; excellence in physics today is essential to leadership in technology tomorrow.
- We must be able to educate the skilled physicists who are needed to carry forward our national programs in energy, the environment, and defense, who can meet the many demands of industry, and who can advise the government on the scientific and technical issues that often underlie urgent policy issues. For our universities and colleges to attract able students and train them at the forefront of knowledge, the quality of research in this nation must be maintained at the highest possible level.

For all these reasons, continued excellence in physics is vital to the United States. The following sections discuss issues that broadly affect our ability to meet the challenges that face physics, and they present recommendations for moving forward.

THE FUNDING PROCESS

Priority recommendations and decisions on the funding of research in physics are made in several stages involving a progressively broader range of scientific and societal considerations. They ultimately encompass political decisions at the level of the President and the Congress based on considerations of national purpose such as national security, economic progress, international competition, national pride, and the distribution of scarce resources.

The nature of the scientific input to the decision-making process depends on the character of the particular research. In some fields the research is performed chiefly by individuals or small groups. These fields advance along many fronts and the scientific priorities are dominantly established by the numbers of researchers who commit themselves to pursue each particular subject area. Further input often comes from disciplinary assessments. In fields that use major facilities, an organized consensus is generally necessary to establish the scientific need for a new facility. Special panels or workshops are usually convened to establish the relative scientific priority of the various facility proposals within each field.

Because decisions on the support of physics involve issues that extend far beyond purely scientific considerations, physicists have an obligation to inform the public and its elected decision makers by explaining the nature of their research, the scientific opportunities, and the roles that each subfield plays in science and on the national scene. Meeting this obligation is a central goal of the Physics Survey.

EDUCATING THE NEXT GENERATION OF PHYSICISTS

This nation's success in educating and training physicists of the next generation depends on the quality of our educational institutions, from kindergarten through graduate school. The quality of education in the United States ultimately reflects our national ideals and the value that our society places on intellectual achievement and the search for new knowledge. Throughout the nation today, there is growing concern about education at all levels.

Primary and Secondary Education

With respect to primary and secondary education, the National Commission on Excellence in Education in the report, *A Nation at Risk: The Imperative for Educational Reform*, summarizes our situation with the following chilling statement:

For the first time in the history of our country the educational skills of one generation will not surpass, will not equal, will not even approach, those of their parents.

The facts on education in science and mathematics are grim. The National Science Foundation report, *Science and Engineering for the 1980s and Beyond*, reveals that the fraction of students who have any contact with physics is so small that we are becoming a nation of scientific illiterates. Our standards for secondary education in science and mathematics are woefully below those of Japan, the Soviet Union, and many of the European countries. The majority of high school physics teachers are underqualified; the supply of qualified new teachers has essentially vanished.

Raising the educational standards in our elementary and high schools presents to the nation a major challenge that demands local, state, and national efforts. Dealing with this complex issue is beyond the scope of the Physics Survey, but we would be negligent not to emphasize the critical nature of the problem and not to endorse efforts to improve secondary education, particularly education in science. In this regard, we welcome the re-establishment by the National Science Foundation of the Directorate in Science and Engineering Education.

Undergraduate Education

Education at the undergraduate level in our colleges and universities is also a matter of concern. A student's undergraduate experience is usually a crucial factor in that student's decision on whether to pursue a career in physics. To achieve scientific excellence, the nation must maintain the highest possible standards in its undergraduate programs, but our colleges and universities face increasing difficulties in doing so. The recent report, *Involvement in Learning: Realizing the Potential of American Higher Education*, prepared by the study group on the Conditions of Excellence in American Higher Education, cites such problems as underpaid faculty, overspecialized curricula, and deteriorating buildings. Physics, as part of the core of higher learning, shares these problems.

Undergraduate training in physics is carried out roughly equally at Ph.D.-granting and non-Ph.D.-granting institutions. At many universities, undergraduate education benefits from the research programs of the faculty. The 4-year colleges, lacking this advantage, are more vulnerable to shrinking enrollments, the loss of financial resources, and the problems arising from decreased interest in undergraduate education at the national level.

The status of undergraduate education in physics is being surveyed by the Committee on Education of the American Physical Society (APS). Preliminary studies reveal the need for concern on such issues as updating and augmenting undergraduate instructional equipment, facilitating faculty and student participation in research, strengthening visiting scientist programs, and encouraging the development of new courses and curricula. There is, for example, a real need for senior-level courses in plasma and fluid physics. The creation of the College Science Instrumentation Program by the National Science Foundation is a welcome step toward enhancing undergraduate education. We urge the colleges, universities, and federal agencies to be responsive to the findings of the APS Committee on Education, as well as to forthcoming studies by other interested groups.

The broad problems of precollege and undergraduate education in the United States deserve serious attention. The discussion here, however, will focus on professional training at the graduate level.

Education at the Graduate Level

We look to our universities to train the physicists who will extend the frontiers of knowledge, carry forward our national programs, and help create new technologies. The scientific and cultural vitality of the universities, the quality of the faculties, and the excitement of the research are all crucial factors in attracting and educating capable young scientists. Because it is essential for the health of physics and because we find it to be in difficulty, university research is a central issue of this report.

Most nations isolate forefront research from their educational institutions; the United States does not. On the contrary, student participation in research at the highest professional level is at the heart of U.S. graduate education. This tradition is widely regarded as a special source of our strength in physics. More than half of this nation's basic research in physics is carried out within the universities. Thus, our universities not only provide an essential educational service but they are in themselves a vital force in research. Because professional training in physics for capable young men and women is essential to the welfare of this nation, and because university research is the largest single element in our basic research effort, maintaining excellence in physics demands that we maintain excellence in our universities.

This overview report provides a natural forum for addressing the issue of research in our universities, which affects every field of physics and every style of research. *The issue needs urgently to be addressed because evidence in the panel reports indicates that the climate in which university research takes place is generally troubled.*

The critical factor underlying any discussion of the climate for research in the universities is the strength of physics itself—the challenges, the opportunities, the sense of excitement and progress that animates science. The panel reports that constitute the main body of this Physics Survey provide evidence of an enormous vitality in physics today. Their descriptions of discoveries in recent years, summarized in the previous chapter, provide an accounting rich in scientific achievement. Scientific opportunities today are abundant.

For the promises of physics to be fulfilled, however, the climate for research in our universities must be healthy. The factors that make for health in research organized around major facilities and national laboratories are somewhat different from the factors important to university-centered research. In fields like elementary-particle physics, which depends on major facilities, the quality of university research is tightly linked to the quality of the facilities and the laboratories that support them. Careful attention must be given to particular issues that affect university research in these areas, including such matters as the special quality of educational experience in a large laboratory, the effect of off-campus (and sometimes out-of-country) research, and career development in large groups. These issues are discussed in the panel reports of each of the relevant subfields. Beyond these special considerations, however, it is essential that, in planning for large facilities and major programs, adequate support be set aside for the university-based component of the research. Otherwise, the nation runs the risk of having the facilities of the future without the physicists to use them.

Most basic research in physics is not carried out by groups working at major facilities, however, but by small groups, which usually work with equipment in their own laboratories. These activities, which we shall call collectively "small-group physics," constitute the backbone of university research. When viewed on a one-by-one basis, they appear as a collection of relatively small and somewhat disconnected research efforts. We believe, however, that it is essential to view small-group physics not one by one but as an entity, because only in this way can one obtain a coherent picture of research in the universities and on the national scene. The following section explains this point of view.

RESEARCH IN SMALL GROUPS

Small-group research encompasses those areas in which the research is generally pursued by a few investigators working together, possibly only a single scientist with a few students, most often using equipment in their own laboratories. Much of condensed-matter physics operates in this style, as do atomic, molecular, and optical physics; fluid physics; and certain areas of astrophysics and nuclear physics. Theoretical physics in many subfields is organized in this mode, as well as most of the research that interfaces with the other sciences (for instance, biophysics and medical physics).

Most areas of small-group physics usually advance by a multitude of discoveries that fit together to reveal a major scientific advance, in contrast to research that is organized around a single conceptual theme. We are, for example, beginning to understand surfaces with the detail and clarity that are characteristic of basic physical theory. Our understanding of two-dimensional structures, of the relation between orderly and chaotic motion, and of the nature of surface dynamics has been dramatically deepened. Applications of this research to the creation of new materials and to catalysis promise to have important influences on industry. In viewing the creation of contemporary surface physics, however, it is not possible to cite the *one* crucial experiment, new technique, or theoretical breakthrough that should be credited for the advances. The progress is due to theoretical advances and to research by many groups using a host of techniques—some highly novel, others traditional.

Achievements of small-group research include such discoveries as spontaneous symmetry breaking and renormalization group theory, the creation of new forms of matter such as clusters and one-dimensional conductors, and rapid progress in the understanding of chaos and turbulence. Because of the relatively small scale of the individual research activities, small-group research is flexible and can move rapidly in response to discoveries. New fields can spring into existence; artificially structured materials and femtosecond spectroscopy are two recent examples. The intellectual challenge and the flexible style of small-group research attract some of the most able physicists, as the high number of Nobel Prizes awarded in these areas attests.

Small-group physics has had a major impact on the nation's economy through generating advanced technologies and new industries. Our modern optics and electronics industries, for example, have their roots in small-group research. Much of the advanced instrumentation now used by industry, science, and medicine has come from these areas.

Research carried out in such groups plays a major role in educating professional physicists. Condensed-matter physics and atomic, molecular, and optical physics train slightly more than half of all the students who receive doctorates in physics; more than 70 percent of the doctorates in the United States are awarded for research in small groups. In fact, the most important aspect of small-group research may well be the opportunities for initiative and innovation provided for research students working in these areas. These aspects are precisely those that must be experienced by young scientists if physics is to continue its rapid intellectual advancement.

In the universities, the various fields of small-group research face a number of similar problems. Foremost is a critical need for laboratory equipment or instrumentation. Inadequate support for instrumentation in the United States was identified as a growing problem in the early 1970s in the previous Physics Survey (*Physics in Perspective*, National Academy of Sciences, Washington, D.C., 1972). The situation has steadily deteriorated since then. The most recent studies, *Revitalizing Laboratory Instrumentation* (National Academy Press, Washington, D.C., 1982) and *Academic Research Equipment in the Physical and Computer Sciences and Engineering* (National Science Foundation, 1984), report that essential instrumentation in university research laboratories is obsolete or simply nonexistent. Lack of up-to-date equipment is cutting off the universities from forefront research. The research groups are losing their ability to compete on an international level, and our students are not being trained in the state-of-the-art techniques needed by industry and government. There is a shortage not only of larger pieces of equipment, such as laser systems, molecular-beam epitaxy machines, and surface-scattering apparatuses (which can cost from a quarter of a million dollars to more than $1 million), but also of equipment such as superconducting solenoids or high-vacuum systems (which can cost $50,000 or more), and even of small instruments like oscilloscopes and signal generators.

The Department of Defense-University Instrumentation Program illustrates the size of the problem. This program, budgeted at $30 million per year for 5 years, received requests totaling more than $645 million for the first year, and this figure represented only a fraction of the total need.

A second problem common to small-group research in the universities is the loss of the infrastructure of support services that are essential in forefront research. Machine shops, electronics shops, and special services such as materials preparation have deteriorated or disappeared from universities across the nation.

The lack of instrumentation and support services is symptomatic of a single underlying problem in small-group research. Since the early 1970s, the base level of support in these fields has lagged far behind the costs of competitive research. There is a widespread misperception of the costs of competitive research in small groups. Equipping a modern experimental laboratory can require over $1 million, though a few hundred thousand dollars is a more typical figure. Operating costs for a healthy university group, including the acquisition of instruments, typically range from $200,000 to $400,000 a year, and some independent groups may require an annual budget of $1 million. In contrast to these costs, the average grant size in many small-group areas is about $80,000 or less a year. This enormous disparity between the size of the grants and the costs of research is making it increasingly difficult to carry research forward.

The most serious impact of underfunding, however, is on the development of young talent. The number of new grants funded across the nation every year is small; a prospective faculty member must face the possibility of having to wait several years to launch a research project. Beyond this difficulty lies the prospect of pursuing a research career in a situation of perpetual shortage. As a result, academic careers have become significantly less attractive than they were in former years. In some areas, the universities are no longer able to compete with industry or government laboratories for the best talent. Unless the climate for research in the universities is significantly improved, we face the possibility of a critical shortage of highly qualified young physicists to fill these positions.

The Panel on Condensed-Matter Physics and the Panel on Atomic, Molecular, and Optical Physics, meeting separately and addressing different research communities, arrived at the same conclusions: the need for instrumentation is urgent, and it is essential to bring the support of the groups up to a realistic level. The panels estimated the costs by somewhat different processes, in one case by estimating broadly across many research activities having different needs, in the other case by carrying out a group-by-group tally of research costs. The results were essentially identical: *to allow a reasonable number of groups to pursue the new scientific opportunities, and to allow some young investigators to enter the field, the level of operating funds must be doubled over about a 4-year period.*

The major fraction of small-group research is carried out in universities, and one can make reasonable estimates of the actual cost of this component of research. The federal expenditure for physics research in the universities in 1983 was $339 million. Approximately $260 million of this total was spent in support of independent-group activities: $150 million for condensed-matter physics and atomic, molecular, and optical physics and $110 million for other areas. In addition, approximately $65 million was spent in the universities for large facilities.

To allow independent-group activity in physics to flourish, the base support of the work in the universities needs to be augmented by $260 million over a 4-year period, or by $70 million per year in 1985 dollars, for each of the 4 years.

When one views small-group research collectively and witnesses its enormous effect on science education and our universities, as well as the abundant returns it yields to society, the endeavor surely represents one of the most important ways for the nation to invest in research.

LARGE FACILITIES AND MAJOR PROGRAMS

Progress in basic and applied experimental physics depends ultimately on progress and innovation in the apparatus and instrumentation used in experiments. Often the great forward strides in physics have been based on the invention of new kinds of physics instruments or on major changes in experimental techniques. Major instrument inventions in the last half century include the laser, the electron microscope, the particle accelerator, and the magnetic confinement apparatus used in plasma fusion studies. Examples of major changes introduced into physics techniques are the use of integrated circuits and high-speed computers, the use of rockets and satellites for atmospheric and space physics, and the use of very low temperatures to study the properties of matter.

Some instruments (lasers and low-temperature equipment, for example) are small and of moderate cost; they can be owned and used by a single laboratory of average size. But other instruments or techniques involve large, complex, and expensive equipment. Examples are particle accelerators (Figure 3.1), magnetic and inertial plasma-confinement apparatuses (Figures 3.2 and 3.3, respectively), and space satellites. Such instruments and techniques require the large facilities and major programs discussed in this section. These facilities, which are located mostly at laboratories and centers supported by the Department of Energy (DOE), play an essential role in physics research.

Large facilities have been developed to allow the physicist to work at the cutting edge of research in many areas of physics. (The types of facilities used by the various subfields are listed in Table 3.1; the costs of some of the proposed new facilities are listed in Table 3.2.) Cutting-edge research offers the highest probability of breaking through into new areas of science and technology and the best promise of answering the deepest questions. Much cutting-edge research can be done with instruments of moderate size and cost, but some aspects of

54 PHYSICS THROUGH THE 1990s: AN OVERVIEW

FIGURE 3.1 The Cornell Electron-Positron Storage Ring (CESR). The diameter is about 800 ft.

FIGURE 3.2 The Tokamak Fusion Test Reactor at the Princeton Plasma Physics Laboratory.

FIGURE 3.3 The Particle-Beam Fusion Accelerator (PBFA) at Sandia National Laboratories. This second-generation machine of its type will generate 30 MV and several hundred terawatts when it is completed in 1986. It will be used for inertial fusion, driven with Li^+ beams.

TABLE 3.1   Main Types of Large Facilities Used in Physics Research

Physics Subfield	Main Types of Large Facilities
Elementary-particle physics	Particle accelerators
	Particle colliders
	Comprehensive particle detectors
Nuclear physics	Particle accelerators
	Particle colliders
	Fission reactors
Condensed-matter physics	Synchrotron radiation sources
	Fission reactor neutron sources
	Pulsed spallation neutron sources
Plasma physics	Magnetic plasma-confinement devices
	Inertial fusion devices
Gravitation, cosmology, and cosmic-ray physics	Gravitational radiation detectors
	Ground-based and space-based telescopes
	Ground-based and space-based cosmic-ray particle detectors

TABLE 3.2   Proposed Large Physics Construction Projects (Currently in Planning and Proposed Stages)

Project	Proposed Starting Date	Total Estimated Cost (Millions of Dollars)
Superconducting Super Collider (SSC) (Elementary-particle physics)	1988	>3000
Burning Core Experiment (BCX) (Plasma physics)	1988	300-500
High Flux Reactor (Condensed-matter physics)	1989	260
Relativistic Heavy Ion Collider (RNC) (Nuclear physics)	1988	250
Continuous Electron Beam Accelerator Facility (CEBAF) (Nuclear physics)	1986	225
6-GeV Synchrotron Facility (Condensed-matter physics)	1987	160
Gravity Probe B Satellite (Gravitational physics/NASA)	1988	120
1-2 GeV Synchrotron Facility (Condensed-matter physics)	1987-1988	90
Gravity-Wave Detector: Laser Interferometer (Gravitational physics/NSF)	1987	50

it require large instruments. For example, frontier research in elementary-particle physics and in much of nuclear physics requires particle accelerators. The discovery of the J/ψ and Υ particles, the discovery of the τ lepton, the experimental demonstration of the unification of the weak and electromagnetic forces culminating in the finding of the W and Z bosons, and the discovery and study of nuclei far from equilibrium all required particle accelerators. The productive studies of matter that use neutron scattering require nuclear reactors or high-power accelerators. And much frontier research in physics, chemistry, and biology requires the intense light and x-ray beams that can only be obtained at synchrotron radiation facilities (Figure 3.4).

In a subfield, the decision whether to build and operate a large facility or to carry out a major program can mean that funds and manpower may not be available for other research. In the last few decades, therefore, as the need for major facilities has become more common in physics, the physics community and the funding agencies have examined more and more closely how a proposed large facility

FIGURE 3.4 The National Synchrotron Light Source at Brookhaven National Laboratory. The Ultraviolet Radiation Synchrotron ring is in the upper central portion of the picture with various experimental parts emanating from it.

will contribute to research. The major criterion has been: "Is this facility necessary to maintain research at the cutting edge?" The evaluations have been carried out by a variety of committees and panels of the National Research Council and the federal government. Where the answer to the above criterion question has been "no," the proposed facility has been rejected. Examples of facilities not funded include the full-scale NOVA laser, the Isabelle project, and the toroidal fusion core experiment.

We summarize below the large facilities and major programs recommended by the panels of the Physics Survey Committee. The recommendations are based on the evaluations described in the previous paragraph as well as on the deliberations of the panels themselves. We again refer the reader to the subfield reports for details on the research that can be performed with these facilities—and only these facilities.

Elementary-Particle Physics

Frontier research in elementary-particle physics requires the production of intense beams of very-high-energy particles by accelerators. High energies are necessary to probe the internal structure of the elementary particles because these particles are held together by very strong forces. High energies are also necessary because by converting energy into mass, the physicist can search for the more massive particles that seem to hold the key to our understanding of the origin of matter itself and the unification of the fundamental forces. Thus a program is recommended whose key elements are (a) the proposed construction of the highest-energy colliding-beam accelerator in the world, the Superconducting Super Collider, and (b) the extension of the high-energy capabilities of two existing accelerators by the addition of collider facilities (Figures 3.5 and 3.6). This program has also been recommended by the DOE High Energy Physics Advisory Panel.

THE SUPERCONDUCTING SUPER COLLIDER

The U.S. elementary-particle physics community is carrying out an intensive research, development, and design program intended to lead to a proposal for a very-high-energy proton-proton collider, the Superconducting Super Collider (SSC). It will be based on the accelerator principles and technology that have been developed at several

FIGURE 3.5 A view of the 2-mile-long accelerator at the Stanford Linear Accelerator Center (SLAC) looking along the tunnel. The accelerator has just been rebuilt to provide 50-GeV electrons and positrons for the Stanford Linear Collider.

national laboratories, including the extensive experience with superconducting magnet systems gained at the Fermi National Accelerator Laboratory (FNAL) and Brookhaven National Laboratory (BNL). The SSC energy could be as high as 40 TeV, providing by far the highest-energy particle collisions in the world. This very high collision energy is needed to search for heavier particles, to answer the question of what generates mass, and to test new theoretical ideas about the fundamental nature of matter, energy, space, and time. Furthermore, history has shown that unexpected discoveries made in a new energy regime often prove to be the most exciting and fundamentally important for the future of the field.

60 *PHYSICS THROUGH THE 1990s: AN OVERVIEW*

FIGURE 3.6 Final construction of the Collider Detector (CDF) at the Fermilab Tevatron Proton-Antiproton Collider. The detector is about 35 ft high. The Tevatron is the highest-energy particle collider in the world.

EXTENSIONS OF THE CAPABILITIES OF EXISTING ACCELERATORS

The capabilities of two existing accelerators in the United States are currently being extended into new areas of elementary-particle research by adding collider facilities to each of them.

• A 100-GeV electron-positron collider, using a new linear collider principle, is now being constructed at the Stanford Linear Accelerator Center (SLAC). This machine will provide high-energy particle collisions that can be studied in a relatively direct way because electrons and positrons are simple particles.

- The Tevatron at FNAL is being modified so that the superconducting ring can also be operated as a 2-TeV proton-antiproton collider. This will provide the highest-energy particle collisions in the world until an accelerator such as the SSC is built.

To maintain the cutting edge in research in the U.S. program in elementary-particle physics, it is essential to complete these additions on schedule.

SUPPORT OF EXISTING AND EXTENDED FACILITIES

Most elementary-particle physics experiments in the United States are carried out at four accelerator laboratories. Two fixed-target proton accelerators are now operating: the 30-GeV Alternating Gradient Synchrotron (AGS) at BNL and the 400- to 1000-GeV superconducting accelerator, the Tevatron, at FNAL. Cornell University operates the electron-positron collider CESR. SLAC operates a 33-GeV fixed-target electron accelerator that also serves as the injector for two electron-positron colliders, SPEAR and PEP. In addition, some elementary-particle physics experiments are carried out at medium-energy accelerators primarily devoted to nuclear physics.

Experimentation at accelerator laboratories requires complex particle detectors that are often major facilities in their own right. These detectors are as crucial as the accelerators themselves. Because of the large fixed costs of accelerator laboratories their productivity can be increased considerably by a modest increase in equipment and operating funds.

Nuclear Physics

The major new frontier of nuclear physics is the investigation of the quark-gluon nature of nuclear matter to attain a deeper level of understanding of the structure and dynamics of atomic nuclei. Two nuclear-physics accelerators, of complementary natures, are recommended for pursuing this goal: one is an electron accelerator, allowing studies of the details of nuclear structure with unprecedented precision; the other is a relativistic heavy-ion collider, making possible studies of nuclear matter in regions of energy density never explored before. Each will be the premier facility of its kind, providing a wide range of cutting-edge research methods in nuclear physics.

THE CONTINUOUS ELECTRON BEAM ACCELERATOR FACILITY

The Continuous Electron Beam Accelerator Facility (CEBAF) is a 100-percent-duty-factor, 4-GeV linear-accelerator stretcher-ring complex. A major research focus of CEBAF will be the investigation of the microscopic quark-gluon aspects of nuclear matter, using the electron beam to probe with high precision the detailed particle dynamics within an entire nucleus. Also studied will be the nature of the transition, in nuclear matter, from the low-energy regime of nucleon-nucleon interactions (best described by independent-particle models of nuclear structure) to the intermediate-energy regime of baryon resonances and meson exchange currents (described by quantum field theories of hadronic interactions in nuclei) and the ensuing transition to the high-energy regime of quarks and gluons (described by quantum chromodynamics).

THE RELATIVISTIC NUCLEAR COLLIDER

The Relativistic Nuclear Collider (RNC) is a variable-energy, relativistic heavy-ion colliding-beam accelerator, with an energy of the order of tens of GeV per nucleon for beams of heavy ions with atomic numbers up to that of uranium. A major research focus of the RNC will be investigations of one of the most striking predictions of quantum chromodynamics: that under conditions of sufficiently high temperature and density in nuclear matter, a phase transition will occur from excited hadronic matter to a quark-gluon plasma, in which the quarks, antiquarks, and gluons of which hadrons are composed become deconfined and are able to move about freely. The quark-gluon plasma is believed to have existed in the first few microseconds after the big bang, and it may exist today in the cores of neutron stars. Producing it in the laboratory would be a major scientific achievement, bringing together various elements of nuclear physics, particle physics, astrophysics, and cosmology. The Nuclear Physics Panel endorses the 1983 Long-Range Plan of the Nuclear Science Advisory Committee (NSAC) in recommending the construction of this accelerator as soon as possible, consistent with the construction of the 4-GeV accelerator discussed above.

EXTENSIONS OF EXISTING FACILITIES

Many of the major questions currently facing nuclear physics, including nuclear astrophysics, point to a number of important scientific opportunities that are beyond the reach of the experimental facilities either in existence or under construction. Extensions of existing facilities are required to provide intense kaon, muon, and neutrino beams of high quality; high-resolution polarized proton beams spanning the energy range from 50 MeV to several GeV; secondary beams of radioactive nuclei; low- and medium-energy antinucleon beams; and a solar neutrino detector sensitive to low-energy neutrinos. Decisions regarding the relative priorities of these options, among others, must be made at the appropriate time.

Basic to the entire nuclear-physics program is an adequate level of funding for equipment and operating costs at existing facilities, both large and small. Only if the accelerators are funded to their full operating potential—including the development of the requisite instruments and detectors—can the nation's investment in their construction be fully realized.

Condensed-Matter Physics

Two types of large facilities provide cutting-edge research opportunities in condensed-matter physics: facilities used for the generation of synchrotron radiation and facilities involved in the generation of low-energy neutrons. In addition, a lesser effort is required for the production of high magnetic fields. The priorities for the facilities have recently been examined closely in the report of the National Research Council's Major Materials Facilities Committee (*Major Facilities for Materials Research and Related Disciplines*, National Academy Press, Washington, D.C., 1984). These needs are briefly summarized here.

SYNCHROTRON RADIATION FACILITIES

Synchrotron radiation provides an intense source of tunable radiation from the ultraviolet to the hard-x-ray region of the photon spectrum. The tremendous intensity of synchrotron radiation sources has made possible new studies of both the structural and the electronic properties of materials. It has allowed such advances as angle-resolved photoemission, where electronic band structure is directly measured; extended x-ray absorption fine structure (EXAFS), where local atomic arrangements are examined; and two-dimensional crystallography, where surfaces and extremely thin films (10 nm) are studied. In order to continue progress in this research, it is essential that

- The current new generation of synchrotron facilities be completed as soon as possible, because the high brightness of these facilities will serve the short-term needs of the next 3 to 5 years.

• Capabilities of advanced wiggler and undulation insertion devices be explored, because of their potential for even higher brightness.

• New insertion devices be implemented at existing facilities and new optical devices be developed in parallel to take advantage of those sources.

Finally, the characteristics of current synchrotrons are not optimal for use with a large number of insertion devices. Consequently, a new synchrotron facility optimized for use of insertion devices should be constructed. The increased brightness of such a synchrotron radiation source would create new opportunities in the studies of photoabsorption, EXAFS and its variant spectroscopies, x-ray scattering, and other techniques. It would also make possible new applications to biology, medicine, and earth science. The major facilities report cited above favored a 6-GeV synchrotron for this purpose because it would cover the region of the spectrum in which most x-ray physics is performed, but it also recommended a 1-2 GeV facility to serve the VUV and XUV communities.

NEUTRON FACILITIES

Neutron scattering offers unique opportunities for studying structures and phase transitions in new exotic materials, in magnetic systems, and in systems of lower dimension, under extreme conditions such as high pressures and low and high temperatures. It is also used to study the structure and dynamics of polymers and macromolecular systems and to make precision measurements, such as the determination of the upper limit to the electric dipole moment of the neutron. In order for the United States to stay at the forefront of this research, cold-neutron guide halls and associated instrumentation need to be established at U.S. reactor facilities. This instrumentation will allow new, very-low-energy, and momentum regimes to be probed.

In recent years, a new type of neutron source has been developed for materials research. These sources, called pulsed spallation sources, have been constructed at Argonne National Laboratory and Los Alamos National Laboratory (LANL). They offer new opportunities to explore condensed-matter physics. The United States should continue to explore the use of pulsed spallation sources by the timely completion of the facility at LANL.

As we look to the future, a new high-flux reactor with an order-of-magnitude increase in intensity over existing facilities will be needed because most neutron experiments are intensity limited. A properly designed reactor could, with such intensity, allow new types of measurements using neutrons—the only probe of matter that can look at excitations with large momentum transfers and relatively low energies. The current facilities were built in the 1960s, and a new reactor would represent their timely replacement and improvement.

HIGH MAGNETIC FIELDS

Magnetic fields above 25 teslas are feasible only in pulsed operation. The United States lags behind Japan and Europe in developing high pulsed fields, which allow the exploration of magnetic-field-induced phases that cannot otherwise be explored. Therefore, we recommend increased efforts to produce high pulsed magnetic fields and enhanced instrumentation at the National Magnet Laboratory.

Plasma Physics

Much of the research in plasma physics is directed toward the goal of controlled thermonuclear fusion in a fusion reactor, leading to the production of energy. Reaching this goal requires the simultaneous achievement of high temperatures, high densities, and long confinement times in plasmas—similar to the plasma conditions at the centers of stars. Two types of large facilities are used to carry out cutting-edge research under these extraordinary conditions: devices for magnetic confinement of plasmas (such as tokamaks and magnetic mirror machines) and inertial fusion devices driven by very-high-power laser beams or ion beams. The scientific feasibility of controlled fusion is likely to be demonstrated in the coming decade, and the program outlined below is designed to accomplish this goal.

MAGNETIC FUSION RESEARCH

In all the main approaches to the magnetic confinement of fusion plasmas, the principal measures of performance—plasma density, temperature, and confinement time—improved by more than an order of magnitude as a result of intensified fusion research in the 1970s. One approach, the tokamak, has already come within a modest factor of meeting the minimum plasma requirements for energy breakeven in deuterium-tritium plasmas. The science of plasma confinement and heating has reached a stage that justifies a vigorous research program in magnetic fusion, with the following principal features:

- A base research program involving moderate-size experimental facilities is essential. The program should emphasize both increased scientific understanding of hot, dense plasmas and research on improved confinement concepts (advanced tokamaks, tandem mirrors, and other approaches). The program goals should be both to increase our knowledge of the physics of plasmas and to improve the prospects for fusion reactors. Historically, the interplay between these two research efforts has led to the most creative physical insights and concepts. Such a program is essential to technical progress and to the education of talented new people.
- The demonstration and experimental study of an ignited fusion plasma is the obvious next research frontier after attainment of the energy breakeven point in a plasma. While the scientific understanding of many key plasma phenomena can best be gained on moderate-size experimental facilities, ultimately plasma-confinement properties must be investigated under conditions of intense alpha-particle heating, which will require an ignited plasma core. Fusion research is at the point where consideration of such experiments can proceed with some degree of realism. Obviously, ideas will continue to evolve rapidly as results from experiments, particularly from the Tokamak Fusion Test Reactor (TFTR), become available over the next several years. In the near future, studies of a burning-core experiment should emphasize maximum scientific output with minimum project cost, in a manner consistent with the recommendations of the Magnetic Fusion Advisory Committee (MFAC).

INERTIAL FUSION RESEARCH

During the past decade, a vigorous research effort has been established to investigate the inertial-confinement approach to fusion. In this approach a pellet is to be driven into fusion by the sudden and intense injection of energy from a laser or particle beam. An impressive array of experimental facilities has been developed; inertial fusion drivers include neodymium-glass and CO_2 lasers and light-ion accelerators. This has led to considerable scientific and technological progress. On the basis of such progress, it is important to implement the following near-term strategy for inertial-confinement fusion research:

- Use present driver facilities to determine the physics and scaling of energy transport and fluid and plasma instabilities to regimes characteristic of high-gain targets.

- Use the new generation of drivers coming into operation to implode deuterium-tritium fuel mixtures up to 1000 times liquid density required for high-gain targets and to implode scale models of high-gain targets to the density and temperature of the full-scale target.
- Identify and develop cost-effective, multimegajoule driver approaches.

Timely execution of this strategy will provide the basis for a decision in the late 1980s on the next generation of experimental facilities. Drivers in excess of a megajoule would allow demonstration of high-gain targets for both military and energy applications.

Space and Astrophysical Plasmas

A broad variety of plasmas exists in outer space, ranging from the hot, dense plasmas in the interiors of some stars to the tenuous plasmas of space itself. Research in space and astrophysical plasmas is correspondingly broad, involving major research programs and space research facilities in other sciences as well as physics. We note here by way of summary that cutting-edge research in this area of plasma physics requires implementation of the comprehensive research strategy outlined in the report of the Committee on Solar and Space Physics of the NRC Space Science Board, *Solar System Space Physics in the 1980's* (National Academy of Sciences, Washington, D.C., 1980). These programs, and especially the International Solar Terrestrial Program, are the primary ones that will explicitly contribute to our knowledge of the physical processes in large-scale plasmas. The comprehensive programs proposed in the report of the NRC Astronomy Survey Committee, *Astronomy and Astrophysics for the 1980s* (National Academy Press, Washington, D.C., 1982), will make significant contributions to many problems in plasma astrophysics.

Gravitation, Cosmology, and Cosmic-Ray Physics

Physicists conduct a broad range of research under the general heading of astrophysics—cosmology, nuclear astrophysics, solar physics, and plasma physics, to name a few. To avoid duplication with the report of the Astronomy Survey Committee, the Physics Survey Committee has concentrated on three research areas: gravitational radiation and general relativity; cosmology, particularly as it relates to elementary-particle physics and gravitation; and cosmic-ray physics.

SEARCH FOR GRAVITATIONAL RADIATION

This fundamental consequence of general relativity has not yet been directly observed, but a variety of astrophysical sources are predicted: supernovae collapse, neutron-star binary coalescence, and black-hole formation. Resonant cryogenic bar detectors are approaching interesting levels of sensitivity in the kilohertz frequency range, and at lower frequencies a laser interferometer with 5-km arms is being considered. The vigorous program of the National Science Foundation in gravitational radiation research is strongly supported, and the Long Baseline Gravitational Wave Facility is strongly endorsed.

RELATIVITY GYROSCOPE EXPERIMENT

This experiment uniquely addresses the important magnetic aspects of general relativity by a precision measurement of the precession rate of a gyroscope in an orbiting satellite. Clearly, this is an exceedingly difficult experiment that is many times more sophisticated than any yet attempted in space. The Space Science Board's Committee on Gravitational Physics (see *Strategy for Space Research in Gravitational Physics in the 1980's*, National Academy Press, Washington, D.C., 1981) has recommended that the National Aeronautics and Space Administration (NASA) attempt this difficult but important experiment, and the recommendation is supported by the Physics Survey Committee.

VIGOROUS SPACE PROGRAM IN ASTROPHYSICS

We are in a period of great excitement in cosmology; our understanding of the physics of diverse cosmological epochs and processes is undergoing fundamental changes. Much of the change is traceable to the highly successful U.S. space program. Besides providing unique observations from satellites, space-inspired technology has greatly enhanced the capabilities of ground-based telescopes. The NASA program is sound and forward looking. The following large facility projects, in various stages of development, will make important contributions to cosmology: Hubble Telescope, Cosmic Background Explorer, Gamma Ray Observatory, Shuttle Infrared Telescope Facility, Advanced X-ray Astronomy Facility, Large Deployable Reflector, and an orbiting antenna for the Very Long Baseline Array.

LONG-DURATION COSMIC-RAY EXPERIMENTS

The development of space exploration has resulted in recent dramatic advances in our understanding of cosmic-ray phenomena. For example, we now know that the most abundant cosmic rays represent a sample of matter significantly different from that of our solar system. The development of a series of long-duration (1 or 2 years) cosmic-ray experiments in space is needed.

GROUND-BASED COSMIC RAYS

The only practical means of observation of the most energetic cosmic rays is through the observation of extensive air showers. The Utah Fly's Eye is a unique and successful facility for air-shower studies, and its exploitation and upgrade merit strong support.

Ground-based observation of the highest-energy gamma rays now reveals sources, probably discrete, of energetic cosmic rays. The potential for development in this young field is great, and the deployment of new dedicated detectors for its studies is endorsed.

NEUTRINO ASTRONOMY

The enigma of the discrepancy between theory and observation of solar neutrinos begs for resolution. New solar-neutrino experiments should be developed.

MANPOWER AND EXCELLENCE

As discussed in detail in Supplement 2, the current demand for physicists and their supply appear to be reasonably balanced but only precariously so. The supply of Ph.D. physicists for industry and government has been sustained only because of the decline in the number of academic positions and the increase in foreign graduate students. Many colleges and universities expanded rapidly in the post-Sputnik era, and their staffs have not yet started to retire at a significant rate. Financial problems and shrinking student enrollments have also reduced the number of appointments for young faculty.

The problem of forecasting the future demand for physicists and their supply is complex. The best current estimate is that it should be possible to maintain the balance between demand and supply through 1991, provided that the decrease of U.S. students does not accelerate, that we can continue to attract a reasonable share of foreign students, and that no special demands occur such as major new national programs. We must also not fail to recognize that the increasing reliance on foreign-born physicists—who now constitute 40 percent of our entering Ph.D. students—is cause for unease.

Starting early in the 1990s, faculty positions will become available at an accelerating rate. At that time the demand for physicists will most probably outstrip the supply. Because actions to increase the supply take at least 5 to 7 years to have an effect, we should take steps now to avoid the possibility of a serious shortage of scientists in the 1990s. Among such steps will be making the pursuit of graduate studies and research more attractive. Our specific recommendations are as follows:

• Increase the number of predoctoral fellowships in physics to help reverse the decline in U.S.-born graduate students. These fellowships represent a visible signal to students that the nation needs scientists and that it recognizes and rewards excellence in science. The number currently awarded—45 to 50 each year—could be doubled.

• Attract young scientists to academic careers to assure the continued vitality of university research and to assure continuity in the teaching of physics. Programs such as Sloan Fellowships, Presidential Young Investigator Awards, and the DOE Outstanding Junior Investigator Program are important in this regard. We encourage government funding agencies to work toward attracting young scientists to universities, and we urge industry to participate in the effort.

• Facilitate the entry of U.S.-trained foreign-born physicists who wish to pursue a career in this country. Given the large number of foreign citizens receiving advanced training in the United States, our immigration laws should be simplified to make it easier for these experts to remain in the United States and to use their skills for the benefit of this country.

• Encourage more women and minorities to become physicists. Women and minorities represent an important reservoir of talent in this nation, and every effort should be made to attract them to careers in physics.

POLICY ISSUES CONNECTED WITH MAINTAINING EXCELLENCE

General policies of the federal government having to do with science and technology affect physics research in the United States and are therefore of importance in determining its future. Here we mention two of these policies that have been of importance in the past several years and state our recommendations regarding them.

Role of Industry and Mission Agencies in Basic Research

Fundamental research in physics is one element of the continuous range of activities that constitute the research and development efforts of this nation. The activities on the frontier of research interact with technology not only by producing new concepts and discoveries but also by using new technology. In fact, basic research is frequently the driving force of new technology. Thus for a nation to maintain its technological strength, it must also maintain its strength in basic research. Those industries and enterprises that are dependent on technology in maintaining their leadership must also contribute to basic research to ensure their future.

These statements are particularly true of military technology and the Department of Defense (DOD). Consequently, the Physics Survey Committee strongly recommends that the DOD restore its investment in long-range fundamental research and strengthen its connections with the research community for the mutual benefit of science and national security. To assure that this nation has a viable defense in the future, a constructive relationship between science and the defense establishment is essential.

The relationship between commercial technology and research is also an important element of the continuum mentioned above. Only a small section of American industry has corporately supported research. The U.S. Government and American industry must create an environment, perhaps through tax incentives, that encourages industrial participation in basic research.

Freedom of International Communication and Exchange

Physics is an international enterprise because physical principles know no national boundaries. Physicists everywhere are eager to share in the stimulating exchange of ideas. Evidence for this can be found in the numerous collaborations originating from international visits by members of universities and laboratories, the large international teams that work together on major experiments, and the success of laboratories like CERN in Geneva that are supported by several nations and operated internationally.

For science to flourish, the international freedom of scientists and the free flow of scientific information must be assured. Attempts to curtail the exchange of basic scientific information will only interfere with the growth of science to the detriment of the United States and all nations.

COMPUTATION AND DATA BASES

Computers

Computers are now used in every area of physics, and their role is steadily growing. Nearly all of today's experiments in physics depend on computers, and many experiments would be impossible without them. They are used to control apparatus, to gather data, and to analyze it. Computers are also widely employed by theorists to carry out calculations far exceeding human capability, thus achieving new orders of precision.

Beyond all these applications, large computers are being increasingly used as numerical laboratories in which complex physical systems can be simulated and studied in ways not possible by experiment. Time-dependent processes, such as the motions of electrons during a chemical reaction, the motions of nucleons during collisions, or the evolution of galactic structure, can be visualized, providing a powerful guide to theory. The transition from order to chaos—one of the most profound problems in contemporary physics—can be observed and studied in systems ranging from a few particles to the turbulence around an aircraft. In such applications, computers are providing a new approach to understanding nature called simulation physics. Neither precisely theoretical nor experimental, this style of physics possesses enormous potential, and it is growing rapidly.

The complete range of computers is needed in physics. Microcomputers, minicomputers, and large supermini machines are part of the standard instrumentation of physics laboratories and are also widely used by theorists. Such machines need to be adequately supported as part of the general instrumentation of physics. Networking of small computers holds promise of vastly expanding the capabilities of experimenters and theorists. It is a rapidly changing development that needs continuous evaluation.

Many areas of research require access to the enormous memories and high computational speeds of supercomputers. Special efforts are needed to provide adequate access to these machines.

The general problem of computer needs in science is being addressed by a number of federal agencies. The DOE has taken steps to provide the scientific community with access to supercomputers at both its fusion center and the new supercomputer facility at Florida State University. The National Science Foundation (NSF) has for many years operated a national center for computing in atmospheric physics. Recently the NSF has launched a $40 million initiative to provide four new supercomputer centers for university researchers. NASA, DOD, and the National Bureau of Standards are also addressing the need for supercomputers in science.

We applaud these initiatives for making supercomputers accessible to scientists, but we also call for continuing attention to the need for access by physicists to a full range of computers.

Data Bases

In almost every field of physics, there is a vital need for the critical evaluation, compilation, and dissemination of data. These data are needed for basic research and for wide areas of applied research in government and industrial laboratories. Different agencies are responsible for maintaining the data bases that provide this essential service in the various subfields. In almost all cases, however, the efforts are understaffed and underfunded. With the rapid experimental advances of the past two decades, the data compilations have often fallen far behind the needs. In this survey, the problem was particularly emphasized by the Panel on Nuclear Physics and the Panel on Atomic, Molecular, and Optical Physics.

When we consider that the actual costs of these services are relatively small, sponsoring agencies should make a determined effort to respond to the need to make the data from physics research widely and readily accessible. The effort should be international in scope and should take advantage of the rapid advances in information technology that have created new tools for the effective dissemination of data.

Supplement 1

International Aspects of Physics: The U.S. Position in the World Community

In the summaries of the progress in physics made in the past decade, and of the challenges and opportunities in physics that lie ahead, the Physics Survey has made little reference to the location of the research. Many of the advances have occurred abroad, and there is every reason to expect that physics abroad will continue to flourish. Because physics is truly international, it is important for us to understand the current position of the United States in the world physics community. It is our intent to provide a perspective on our position with respect to the European community and to the Soviet Union and Japan, to discuss the increasingly important international aspects of science, and to review our contributions to the education and training of foreign scientists, especially those from less-developed nations.

No nation that aspires to a continuing leadership role in the political, cultural, and economic arenas of the world can forgo the effort to mount forefront research programs in physics, a discipline central to the sciences and a vital source of new technology. Nevertheless, the spiraling costs of research and limitations on the available resources preclude our achieving absolute leadership in all fields. Thus, while we sustain important research programs and launch new ones, we must also strengthen areas of international cooperation between scientists. Central to this effort is our responsibility to encourage the free flow of information and the unimpeded movement of scientists across national boundaries.

In the years immediately following World War II, the United States found itself in a strong position with respect to science abroad, particularly in physics. Under the threat of war during the previous decade, many physicists—some very distinguished—had emigrated to the United States. The trend continued unabated through the 1940s. The war devastated the manpower and the infrastructure of science in Western Europe, the Soviet Union, and Japan. One consequence of this was a scientific isolation abroad that persisted into the postwar period.

In the United States, however, the war led to greater recognition of the importance of physics, primarily because of the profound changes wrought by the harnessing of nuclear fission for weapons, for energy production, and for research. The disparity between the ample resources and support for science in this nation and those available abroad provided an enormous advantage for the development of physics in the United States in the two decades following the war. Throughout this period, Western Europe, the Soviet Union, and Japan were still in a recovery phase.

A striking indication of the dominant position that the United States attained in world science during these two decades is that English became the accepted language of scientific communication. It is now the standard language at international conferences, and it is the language in which many major foreign scientific journals are now published.

More than four decades have elapsed since World War II, and it is now evident that the other major industrial nations of the world, particularly West Germany, France, Japan, and the Soviet Union, have resumed their rightful places as leaders in the international scientific community. As their economies prospered, these nations invested in education, experimental facilities, and new institutes that, in many cases, exceeded in quality and size the comparable efforts being made here.

There is every reason to believe that these nations will continue to aspire to leadership in physics. Our standing in the international physics community a decade hence will depend in large measure on the priorities we set for ourselves today and on the investment that this nation makes in science and science education in the years ahead.

EXPENDITURES FOR SCIENTIFIC RESEARCH IN THE UNITED STATES AND ABROAD

General Trends

Comparisons of support for basic physics research in this country and abroad are difficult to make because, with some exceptions, data from other countries are either not available or are not separated from general research and development (R&D) expenditures. Nevertheless, the R&D expenditures provide a qualitative indication of trends that apply across all areas of science.

One measure of the relative R&D expenditures for various countries for the years 1961-1983 is obtained by normalizing these expenditures as a percentage of the gross national product (GNP). (See Figure S1.1. Note that Figure S1.1 includes defense and space as well as civilian R&D.) The United States outspent France, West Germany, and Japan combined ($48 billion to $46.3 billion) in the late 1970s. However, the fraction of the GNP devoted to R&D here compared with that of West Germany, Japan, and the Soviet Union declined during this period. A possibly more useful way to identify the trends for support of basic science is to compare the estimated ratio of civilian R&D expenditures to GNP; these are shown in Figure S1.2 for selected countries. It is clear that West Germany and Japan—our two most successful economic competitors—not only surpass the United States but they are rapidly increas-

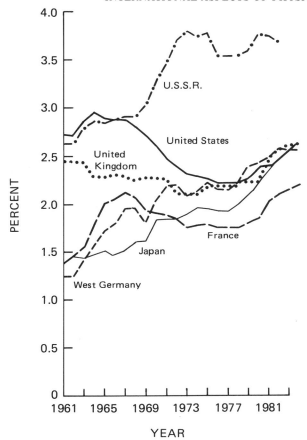

FIGURE S1.1 National expenditures for performance of R&D as a percentage of GNP by country.

ing their lead by encouraging R&D to grow faster than their national economies. A substantially larger fraction of civilian R&D is devoted to energy in Western Europe and Japan, presumably because of their higher energy costs and almost total dependence on foreign oil and coal.

One must exercise care in drawing what might seem obvious conclusions from these data. No one knows what, if there is one, is the optimum ratio of expenditures for R&D to GNP, nor can one readily attribute notable research accomplishments or successful technological applications to absolute expen-

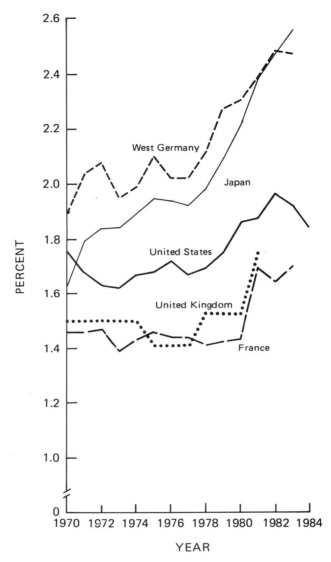

FIGURE S1.2 Estimated ratio of "civilian" R&D expenditures to GNP for selected countries.

diture levels. Optimal support is linked to the effective use of available sources at one's command, the supply of properly trained scientists and engineers, a wise and responsive science policy, and, most important, the challenges and opportunities that fundamentally motivate the expenditures.

Although time lags of 5 to 10 years are not uncommon between R&D expenditures and research and technological realizations, the general trends of the past two decades are clear. They indicate a rapidly growing base of support in Western Europe and Japan, relative to our own, which portrays an increasingly competitive posture on their part in all aspects of science and technology and, one can infer, in their economic development.

Trends in Specific Areas of Physics

Nuclear physics and elementary-particle physics are two areas in which one can meaningfully compare research expenditures in the United States with those in other countries. In the United States, essentially all funding for research in these two fields comes from the Department of Energy (DOE) and the National Science Foundation (NSF) and goes toward the support of national and university-based laboratories and university user groups. Thus, relatively precise information is available from primary funding sources. Abroad, likewise, nearly all funding for research in nuclear and elementary-particle physics is of direct governmental origin.

Comparisons of absolute expenditures can be misleading because they do not take into account such factors as the relative numbers of participating scientists and differing rates of inflation. Again, it is probably more relevant to normalize expenditures to the GNP or the per capita GNP. For basic nuclear research, this comparison for Canada, Belgium, Germany, France, Italy, the Netherlands, Switzerland, and the United Kingdom is shown in Figure S1.3 for 1982. As a percentage of the GNP, the U.S. investment in nuclear research is lower than all but that of the United Kingdom, and significantly less than that of those nations with a per capita GNP in excess of $8000 a year. These figures are consistent with the observations of American physicists who frequently find that nuclear research laboratories abroad are better equipped and staffed than those at home.

A study has been made of the comparative total funding history for elementary-particle physics as a percentage of the GNP in Western Europe, Japan, and the United States (see Figure S1.4). The commitment of the 12 member nations to the support of Conseil European de Recherche Nucleaire (CERN) is demonstrated by the fact that their expenditures have remained roughly constant, whereas ours noticeably declined during the 1970s. Perhaps most striking is the rate of growth of the Japanese effort in this field during this decade; its expenditure/GNP ratio is rapidly approaching our own.

The relative trends in support of atomic, molecular, and optical (AMO) physics and in condensed-matter (CM) physics are more difficult to quantify because of the diverse nature of these fields and their sources of funds. In AMO physics, a marked decline in basic research support at American universities occurred when the Department of Defense (DOD) agencies withdrew much of their support in the early 1970s. In subsequent years little of the lost support was made up for by other agencies. At the same time, these areas were vigorously supported in Europe, particularly in West Germany and France. Consequently, the strength of the European effort relative to ours has grown significantly over the past decade.

Condensed-matter physics is an area of the much broader field of materials science. There is a direct interplay between basic and applied research in this field, with significant support coming from both the private and governmental sectors. The United States has held a commanding lead in this field, in no small part owing to the large and scientifically impressive efforts of industrial research laboratories, notably Bell and IBM. Although no comparable industrial research laboratories exist abroad, increasing emphasis is being placed on materials science, including CM physics, in nationally funded laboratories in Japan and Western Europe. As an example, a major effort in surface science is being mounted at the kernforschungsanlage in Jülich, West Germany, and another in high magnetic fields at the Institute for Solid State Physics in Tokyo. There has been a slow growth in the support base for materials science in the United States, but it is widely perceived to fall short of the rate at which it has grown in Western Europe and Japan.

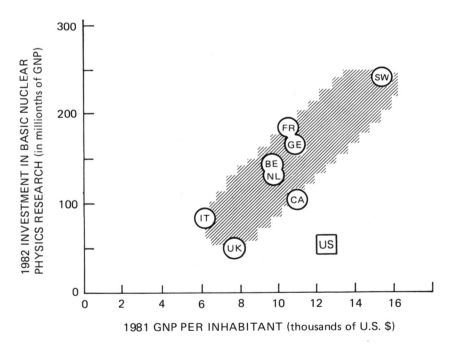

FIGURE S1.3 Comparison of investment in basic nuclear research in other countries with that of the United States for 1982. "Other countries" are Italy, the United Kingdom, Belgium, The Netherlands, West Germany, Switzerland, and Canada.

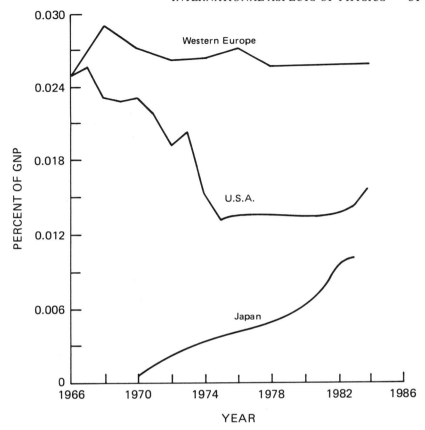

FIGURE S1.4 Funding for high-energy physics as a percentage of the GNP.

THE U.S. POSITION IN BASIC PHYSICS RESEARCH

In attempting to assess the standing of the United States in physics research relative to other countries, one must first establish a set of criteria by which reasonable comparisons can be made. One test might be the *quantity* of research that has been produced per year during the past decade, which can be reasonably reliably measured by the total number of research publications. An assessment of *quality* could be derived from the distribution of international prize awards, such as the Nobel Prize in Physics. Another measure, more controversial, would be to try to identify the most significant advances in recent years in each of the subdiscipline areas. Lastly, a concrete indication of our position in the world physics community can be obtained by considering the major research centers that have been created and the contemplated plans for new ones, both here and abroad. This aspect—the "physical plant"—in which a substantial portion of the research is pursued in fields such as elementary-particle, nuclear, and plasma physics, is documented in the panel reports for the various subfields, and we will not pursue the matter further here. However, it must be realized that the successes of the past and the success that we can hope for in the future are intimately tied to the existence of forefront research facilities, especially in the fields just mentioned.

The most easily measured factor is the number of publications. All significant research is published in the journals of the physics societies in the various countries, or in specialty journals produced by the leading commercial scientific publishers. In the years 1973 through 1980, the percentage of all physics publications worldwide that originated in the United States was as follows*:

Year	1973	1975	1977	1978	1979	1980
%	33	32	30	31	30	30

Some clarifying remarks concerning these figures are needed. First, multiple authorship of a single publication often results from collaborations that are international in nature. In experimental elementary-particle physics, this is the rule rather than the exception, and it is becoming increasingly true in other areas of physics as well. Hence, attaching a quantitative measure based on national origins to particular research publications involves some ambiguity. With this disclaimer, it is nevertheless significant to note that the overall percentage of the research publications in physics that originated in the United States is more remarkable for its size and relative constancy than it is for the slight decline that occurred during the past decade (33 percent in 1973 to 30 percent in 1980). That this is the case, despite the constantly growing proportionate investment being made abroad in R&D relative to our own, attests to the 5- to 10-year lag in realization of the fruits of research investments. It should also be noted that the relative production of our physics research over the same period roughly parallels that in other scientific fields such as chemistry and biology.

Less easily measured is the *quality* of research. Originality, the depth of the discovery, and the probable effect on the future course of scientific thought are some of the qualities that we should consider in judging the importance of a particular research advance. In terms of awards given to recognize major breakthroughs in physics, none is more highly regarded than the Nobel Prize awarded annually by the Swedish Academy of Sciences. The recipients of the Nobel Prize in Physics in the past two decades, along with their countries of residence, are shown in Table S1.1. It is clear that the United States has done extremely well in terms of number of recipients of this distinguished award. At the same time, we must be aware that recognition was bestowed from 5 to 20 years after the work was completed. Furthermore, in the case of experimental research, the facilities needed to accomplish the prized research were often planned and funded years before execution. Thus pride in our recent achievements is no reason for complacency with respect to the future.

Trying to judge what are thought to be the most significant developments in

---

* Science Indicators, 1982; National Science Board (NSF), 1983.

TABLE S1.1 Recipients of the Nobel Prize in Physics (1963-1985)

Year	Recipient	Country
1963	M. Goeppert-Mayer	U.S.A.
	J. H. D. Jensen	Federal Republic of Germany
	E. P. Wigner	U.S.A.
1964	N. G. Basov	U.S.S.R.
	A. M. Prokhorov	U.S.S.R.
	C. H. Townes	U.S.A.
1965	R. P. Feynman	U.S.A.
	J. S. Schwinger	U.S.A.
	S. Tomonaga	Japan
1966	A. Kastler	France
1967	H. A. Bethe	U.S.A.
1968	L. W. Alvarez	U.S.A.
1969	M. Gell-Mann	U.S.A.
1970	L. Néel	France
	H. Alfvén	Sweden
1971	D. Gabor	Great Britain
1972	J. Bardeen	U.S.A.
	L. N. Cooper	U.S.A.
	J. R. Schrieffer	U.S.A.
1973	I. Giaever	U.S.A.
	L. Esaki	Japan
	B. D. Josephson	Great Britain
1974	M. Ryle	Great Britain
	A. Hewish	Great Britain
1975	J. Rainwater	U.S.A.
	B. Mottelson	Denmark
	A. Bohr	Denmark
1976	B. Richter	U.S.A.
	S. C. C. Ting	U.S.A.
1977	J. H. Van Vleck	U.S.A.
	P. W. Anderson	U.S.A.
	N. F. Mott	Great Britain
1978	P. Kapitsa	U.S.S.R.
	A. Penzias	U.S.A.
	R. Wilson	U.S.A.
1979	S. Weinberg	U.S.A.
	S. L. Glashow	U.S.A.
	A. Salam	Pakistan
1980	J. W. Cronin	U.S.A.
	V. L. Fitch	U.S.A.
1981	N. Bloembergen	U.S.A.
	A. Schawlow	U.S.A
	K. M. Siegbahn	Sweden
1982	K. G. Wilson	U.S.A.
1983	S. Chandrasekhar	U.S.A.
	W. A. Fowler	U.S.A.
1984	C. Rubbia	CERN (Italy)
	S. van der Meer	CERN (The Netherlands)
1985	K. von Klitzing	Federal Republic of Germany

the various subfields of physics in recent times is a difficult task. Here one comes closer to the forefront, and the influence of the discoveries is, in some cases, yet to be realized. Taste and style, the numbers of researchers involved in the work in particular areas, and the relative merits of different approaches to the same problem, all play a role in any attempt to judge the significance of research.

If we consider just two fields—elementary-particle physics and condensed-matter physics—we might hope to find a reasonable consensus as to which are the most significant developments in the past decade. For example, in particle theory one might number the following among the 10 important achievements:

1. Asymptotic freedom
2. Grand Unified Theories of weak, electromagnetic, and strong interactions
3. Monopole solutions of gauge theory equations
4. Supersymmetry
5. Generation of baryon number asymmetry
6. Inflationary universe
7. Bag models of hadrons
8. Semiclassical methods in field theory
9. Lattice gauge theory
10. Kobayashi-Maskawa model of CP violation

All but 3, 4, and 10 originated in work begun in the United States.

For the past decade, any listing of major experimental advances in elementary-particle physics would include

1. Discovery and study of the charm quark at SLAC and BNL (U.S.)
2. Discovery of the τ lepton at SLAC (U.S.)
3. Discovery of parity-violating neutral currents in electron-proton scattering at SLAC (U.S.)
4. Discovery of gluon jets at DESY (West Germany)
5. Discovery of the bottom quark at FNL (U.S.)
6. Study of the bottom quark at Cornell (U.S.) and DESY (West Germany)
7. Most recent discovery of the W and Z intermediate bosons at CERN (Europe)

One might add to this the "nondiscovery" of proton decay, work that was performed in the United States and is of fundamental importance to Grand Unified Theories. It must, however, be emphasized again that, with few exceptions, all elementary-particle physics experiments are multinational in conception and execution.

The possible choices for the major advances in condensed-matter physics since 1970 would make a very long list. Certainly, among them would be these:

1. Renormalization group techniques and their application to critical phenomena
2. Superfluid phases of ^{3}He
3. Organic conductors
4. Localization and disorder
5. Normal and fractional quantized Hall effect
6. Charge density waves

7. Artificially structured materials
8. Chaotic phenomena in space or time
9. Effects of reduced dimensionality
10. Valence fluctuations

Almost without exception, in each of the above instances research was started in the United States, or major efforts were mounted here immediately following their initiation elsewhere.

These examples from the fields of elementary-particle physics and condensed-matter physics provide convincing evidence that the United States has been in the forefront of major advances in physics during the past decade or so. There is every reason to believe that we can continue to make outstanding contributions in the years ahead, but we must be willing to invest the resources that are needed to sustain forefront research.

INTERNATIONAL COMPETITION AND COOPERATION

We have examined our position in the international physics community with respect to support trends and our relative standing in recent achievements. While these approaches are useful in establishing benchmarks for overall policy setting, they belie the growing needs for, and trends toward, the increasing internationalization of science. Thus, while arguing that we should maintain a competitive position in all the major subfields of physics, as is appropriate to a leading scientific and technological society, we must also recognize the need for increasing emphasis on supporting efforts aimed at international cooperation.

Increased Internationalization of the Physics Community

Physics, like all science, is fundamentally international; ideas and talent know no national boundaries. Because new research is disseminated at tremendous speed today, every new development in theory and experiment is quickly internationalized. Communities of interest that transcend national barriers are scattered throughout all the fields of physics.

International alliances in physics are not primarily facilities-driven; fundamentally, it is the commonality of interest that generates and nurtures them. The NSF-funded Institute for Theoretical Physics in Santa Barbara, California, provides an illustration. Every effort is made to bring together world leaders in the forefront areas by means of extended workshops. At any time, as many as half of the physicists in residence are from abroad.

Scale and Costs

In some of the forefront areas of physics, the facilities and instrumentation required are so costly as to be beyond the resources of all but the largest and wealthiest nations. Clearly, the accelerators of elementary-particle physics and nuclear physics fall into this category, but so do the tokamaks used in fusion research in plasma physics and the neutron-scattering and synchrotron-radiation facilities used in condensed-matter physics. The most successful example of international collaboration in physics is CERN—the high-energy physics facilities in Geneva, representing a 12-nation West European collaborative venture. CERN is an undoubted scientific success and a monument to the benefits of international collaboration. Another institution that merits recognition is the neutron research reactor facility at the Institute Laue Langevin in Grenoble, France, the largest one of its kind. It is entirely funded by France, West Germany, and Great Britain. Each of these facilities is second to none in the world, and the realization of either of them would have clearly taxed the capabilities of any one of the participating nations if they had attempted it alone.

Avoiding Duplication

Closely related to the issues discussed above is the question of how to use finite resources of manpower and funding most efficiently. Fusion-oriented research exemplifies this issue. All the major industrial nations have good reasons for developing a workable system to extract power from the fusion process. Fundamental research has made substantial progress in the past decade with respect to plasma confinement and heating. Each test facility, however, whether it employs magnetic or inertial confinement, represents a major investment in scientific manpower and money. The United States is engaged in a broad program, allowing for alternative fusion concepts. Nevertheless, we have benefited from extensive exchanges with West European and Soviet scientists on the planning of facilities. The exchanges help to minimize duplication of effort in a field where costs are high and time needed to erect facilities is long.

In elementary-particle physics there is also a growing effort to avoid duplication of research. Future accelerators are being planned to complement existing facilities in other countries, extending the range of energies and the interactions that can be probed. A true regard for the international character of science and a willingness to collaborate at the deepest level with scientists abroad who share a common interest are essential to this planning.

Maintaining Breadth and Depth in Forefront Areas

No nation can excel in every area of physics. From cooperative ventures with other countries, we can benefit in areas where they have special expertise or unusual facilities; and others can benefit from interactions with scientists at our institutions. Notable examples are the collaborative exchange programs between the United States and France, Japan, and Brazil. These programs are jointly sponsored by the National Science Foundation and the corresponding organizations in the affiliated nations. Such collaborations enable U.S. physicists to use and participate in major research with the facilities at CERN and the Institute Laue Langevin. Physicists from abroad might choose to pursue research at FNAL or SLAC.

It is noteworthy that, during the single year 1980, approximately 200 foreign physicists participated in collaborative research at SLAC for periods longer than 3 months. For small nations with limited scientific manpower and other resources, the opportunity to work at a major institution, whether it be ANL, BNL, or Bell Laboratories, may be the only way that they can maintain breadth and depth in physics.

The scientific, technological, and educational programs of some countries with smaller GNP are designed to attract scientific talent from the nations with larger GNP. The Netherlands, Australia, and South Africa provide examples of this type of planning. In attracting foreign scientists to their universities, they have enriched their own institutions while at the same time countering some of the loss due to their own scientists' going abroad.

The United States, though not overtly seeking physicists from abroad, has nevertheless been a magnet. In addition to the outstanding scientific resources here, the international character of our scientific leadership in the post-World War II period made the United States attractive to physicists from abroad. The result was a brain drain from the Western European nations that has abated only in recent years. As those nations continue to strengthen their scientific institutions, however, the net flow of talent may well reverse itself.

FREEDOM FOR SCIENTISTS AND THE FREE FLOW OF INFORMATION

International cooperation presents difficulties as well as rewards (see *Energizing Issues in Science and Technology, 1982*, NSF). For science to flourish, scientists must be free to communicate freely and to move freely. Any interference with these basic principles is a loss for science, a loss for the offending nation, and a loss for the dignity of mankind.

In recent times, there has been an increasing tendency to regard certain scientific and technical information as "privileged." Attempts have been made to restrict or prevent its flow to our political adversaries by means that fall short of actual classification. The ultimate objective of such measures is to slow down the acquisition of our technology by those with whom we are currently at political odds. However, attempts to impede the dissemination of scientific information will inevitably impede our own progress. Scientific secrets are not state secrets; they are held by nature. Our adversaries are as free to try to learn them as we are, without violating national security.

We all share the concern that unfriendly governments have acquired our technology at a rate that some regard as alarming. To stem this outflow, however, our government has taken or is contemplating measures that could be detrimental to the very system that has given us our lead in science and technology. It is the judgment of those who have studied this complex matter that national security is best served by a policy that stresses scientific and technical accomplishments rather than curbs on the free flow of information. (See *Scientific Communication and National Security*, National Academy Press, Washington, D.C., 1982.) There will be narrow, gray areas in which restraints on dissemination will be warranted, but they should be the exception to the rule of free exchange of scientific ideas.

EDUCATION OF FOREIGN PHYSICISTS IN THE UNITED STATES

A major contribution that the United States makes to the world scientific and technological community, albeit a contribution often overlooked, is the education and training of foreign students in science and engineering. Data are available on the numbers of foreign students enrolled in graduate physics programs in the United States: for the academic year 1982-1983, the numbers and countries of origin are given in Table S1.2. Among first-year graduate students in Ph.D. programs, 885 were foreign, as opposed to 1407 U.S. citizens; hence 39 percent of our first-year Ph.D. students were from abroad. Thus foreign students now constitute a significant fraction of our graduate student population.

In 1981, 26.1 percent of the doctoral degrees in physics and astronomy were awarded to foreign students. This represents a substantial increase over that of a decade earlier, when the proportion was 18.6 percent. The situation with respect to physics postdoctorates is more striking yet, particularly in recent years. From 1977 to 1982, the total number of foreign postdoctorals grew from 462 to 672, while U.S. postdoctorals declined from 907 to 652. The majority of postdoctoral physicists in the United States are now foreign. On the one hand, we are fortunate to attract this population of young scientists who play an essential role in carrying forward our research; on the other hand, the increasing reliance on foreign scientists raises serious questions about the number of U.S. students that we are training.

The cost to the United States to educate foreign physicists may be roughly estimated. A yearly expenditure of approximately $15,000 to train each individual is not unreasonable, when averaged over different experimental and theoretical programs. For the approximately 3000 foreign graduate students or postdoctoral associates, the United States thus makes a $45 million annual contribution to the advanced education and training of physicists from abroad. If one also takes into account the 10 to 20 times larger enrollment in undergraduate engineering programs—each of which has a substantial physics teaching component—the total contribution to foreign education in physics and physics training of engineers probably exceeds $70 million each year.

In considering these factors, it is important to note that the United States benefits both directly and indirectly from the flux through its institutions of foreign students and postdoctorates. Many of the young scientists from abroad are among the intellectual elite of their countries; these scientists provide strength and diversity to our programs through their mutual interactions with our physicists. During periods in which too few of our students enter physics as a profession, foreign graduate students and postdoctoral associates often elect to remain here and fill the needs of educational, industrial, and governmental institutions. The United States remains in many ways a nation of great and varied opportunities for scientists from abroad, particularly for those from countries with less-developed scientific establishments.

Most foreign students and postdoctoral associates, however, do not remain in the United States permanently, and for that reason one might question the significant expenditures we make on their education and training. By any measure, our nation remains one of the most advanced in physics education and training. Unquestionably, then, we share the responsibility with the other developed nations to make good use of the opportunities that we can offer. The opportunities are substantial. Ninety-five percent of the world's new science is produced by only 25 percent of the countries of the world. Unless the talents of capable individuals in the underdeveloped nations can be effectively used, the bases for creating technological changes in these nations will not be realized.

The education and training of scientists from less-developed countries, with

TABLE S1.2 Foreign Students Enrolled in Graduate Physics Programs in the United States (Academic Year 1982-1983)

UNITED STATES	4208	LATIN AMERICA	171[a]
		Argentina	19
ASIA	1200	Brazil	21
India	335	Colombia	17
Taiwan, Hong Kong	252	Venezuela	22
China	347	Caribbean Nations	8
Korea	144	Mexico	36
Japan	44	Other Latin America	48
Far East	66		
Other Asia	12	AFRICA	38
		Nigeria	13
EUROPE	338	South Africa	6
Britain	41	Other Africa	19
Austria, Switzerland	16		
Benelux Countries	17	CANADA	70
Germany	51		
Greece	58	AUSTRALIA	17
Italy	32		
Scandinavia	27	OTHER	5
Spain	8		
Eastern Europe	56		
Other Europe	32		
MIDDLE EAST	226		
Israel	13		
Egypt	21	TOTAL ANSWERING	6273
North Africa	26		
Iran	86		
Mideastern Countries	80		

SOURCE: American Institute of Physics.

[a] Total enrollment, including part-time students, is known to be 10,500 for 1982-1983. The number above (6273) was obtained from responses to a questionnaire sent out by AIP.

the expectation that they will return home to staff their research and education institutions, is cost-effective by any measure. If we were to elect to donate the money that it costs to educate foreign physicists here, it is unlikely that we could find any other way in which it could be used as quickly and efficiently to raise the scientific and technological level in the recipient's home country.

Greece is an example of a small country where a sizable effort is being made to raise the level of scientific teaching and research, with the aim of ultimately strengthening the country's industry and economy. About three quarters of the scientists at universities and research centers there received their doctorates at

American universities. It is difficult to imagine how the United States could have bestowed a more valuable gift.

SUMMARY

The United States dominated the world of physics in the decades following World War II. The nations of Western Europe, the Soviet Union, and Japan have now emerged from the protracted recovery phase for science following that war and have resumed their rightful places as leaders in the scientific community.

The United States remains at the forefront in physics, as judged by scientific achievements in that period. Among the indicators are these: the United States is responsible for about 30 percent of the world's research publications in physics; the number of Nobel Prize winners has been high; analyses of major achievements in two fields, elementary-particle physics and condensed-matter physics, reveal that many originated in the United States.

There is a growing need for international cooperation, and in response to this need physics is becoming increasingly internationalized. The great expense of major research facilities increasingly demands international cooperation to secure adequate support, and the expense of some programs, such as fusion, requires international cooperation to avoid duplication of effort.

For international cooperation to be effective, there must be freedom to cooperate. Scientists must be free to travel and to collaborate with other scientists, and the flow of scientific information must not be hindered except for the most urgent reasons.

Foreign students now constitute a significant portion of our graduate-school population in physics. In 1983, 39 percent of our entering graduate students in physics were foreign citizens, and more than half of our postdoctoral workers were born abroad.

The United States makes a major contribution to the less-developed and underdeveloped countries by training scientists who will return home to teach and carry out research. It would be difficult to find any other way that is as cost-effective for raising the scientific and technological level in those countries.

Supplement 2

Education and Supply of Physicists

Numerous opportunities for major advances in physics during the coming decades are described elsewhere in this overview volume and throughout the panel reports. International cooperation and competition can be expected to continue to stimulate the scientific enterprise, but maintenance of a U.S. leadership role in physics is essential to the vitality of physics everywhere and to the technological strength of our nation. Such leadership is possible only with a strong base of highly educated scientists. This supplement describes the changing patterns in the supply and employment of physicists in the United States and poses questions concerning our manpower resources in the future.

The events of the past two decades have left us with an aging academic community, a smaller core of physicists regularly engaged in basic research, and a continuing high outflow of physics Ph.D.s to related areas of science and engineering. Although graduate enrollments in physics began to increase in the 1980s, most of the increase reflected the rapid growth of the foreign-student component. The number of U.S. physics graduate students has continued to decline slowly. The decline, coupled with a low retention of Ph.D.s in physics, leads us to expect only minor increases in the physics labor force in the near future.

By the middle to late 1990s, the retirement rate is expected to increase significantly as a result of the large number of entries in the 1960s. *The supply of entrants into the physics labor force could decline at the very time that retirements will be most numerous.* Although pleas for a return to the high production levels of graduates during the late 1960s would be inappropriate, concern over the effects of a potentially diminishing labor force is warranted.

In the following sections we describe the major features of the current situation and the issues on which they focus attention, including an analysis of the contributory events of the past two decades. We conclude the supplement by projecting the contour of physics manpower in the coming decades. Because the production of highly skilled physicists is a long-term process calling for extensive lead time, immediate consideration of such projections takes on added importance.

PRODUCING TRAINED YOUNG PHYSICISTS—A HISTORICAL OVERVIEW

Initial decisions affecting (or precluding) later careers in science are typically made in the early teen years. Secondary-school students who fail to take the advanced science and mathematics sequence are unlikely to take college-level science courses, and they are even less likely to major in physics. In the United States, however, less than half of secondary-school students take more than 2 years of science and mathematics. A much smaller proportion (less than 20 percent of 1980 graduating seniors) is exposed to physics. The value of even that limited exposure is being questioned in light of the shortage of qualified mathematics and science teachers. Many of the competent teachers have left for more prestigious and remunerative jobs, while the number of new science teachers being trained has declined severely.

The deteriorating condition of precollege science and mathematics education has long been of concern to the physics community, and by the 1980s it had become a national concern.* The crisis in secondary-school education raises serious questions about the future scientific literacy of the nation and the development of the needed pool of potential scientists.

Training professional physicists is a lengthy and selective process. Only a small fraction of the students exposed to physics at the college level major in physics. A still smaller fraction undertake graduate study and emerge 4 to 9 years later with a Ph.D. in physics. This core of Ph.D. physicists is a precious national resource. It is the group that must sustain our physics research effort; train the next generation of students, researchers, and teachers; and provide the talent for a wide variety of related scientific and engineering disciplines.

The number of physics graduate students soared during the 1960s, following a rekindled interest in the physical sciences brought on by Sputnik and increased federal support. More than 1500 Ph.D.s were graduated in each of the last 3 years of the decade. In 1969-1970, Ph.D. production peaked at 1545 (Figure S2.1), a 148 percent increase since the beginning of the decade.†

---

* See, for example, *A Nation at Risk : The Imperative for Educational Reform*, The National Commission on Excellence in Education (1983), and *Educating Americans for the 21st Century*, The National Science Board Commission on Precollege Education in Mathematics, Science and Technology (National Science Foundation, Washington, D.C., 1983).

† Data are derived from American Institute of Physics' (AIP's) annual departmental surveys of enrollments and degrees. Figures may differ slightly from those obtained from other sources, e.g., National Academy of Sciences and National Science Foundation. Nearly 100 percent of physics departments have traditionally responded to these annual AIP surveys. Data missing from some departments are estimated on the basis of the previous year's responses.

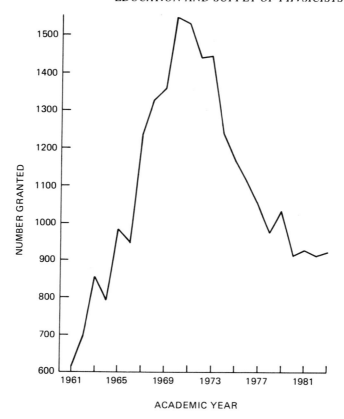

FIGURE S2.1 Physics doctoral degrees granted, 1961-1983.

Employment opportunities in academe were bright during the early 1960s because of the upsurge in student enrollments and the increased federal grant money for research. The number of positions in research and teaching expanded rapidly, and new physics Ph.D.s quickly filled the assistant professor ranks. Although it was difficult to believe at the time, such a high rate of expansion could hardly be expected to continue indefinitely.* It did not. Toward the end of the decade, when the physics enrollment was approaching its peak size, harbingers of difficulties ahead began to appear. Federal funding for basic research slowed with the nation's heightened involvement in the

---

* Physics Survey Committee, *Physics: Survey and Outlook*, National Academy of Sciences—National Research Council, Washington, D.C., 1966.

Vietnam War.* The Mansfield Amendment had the effect of curbing basic research support from many of the traditional mission agencies, and the National Science Foundation was unable to assume the full burden of support. New academic positions diminished in number, and the economy (which had not been particularly robust) presented only limited alternative job opportunities.

By the time the large new classes of physics Ph.D.s entered the employment marketplace at the beginning of the 1970s, traditional opportunities had nearly vanished. Many Ph.D.s took temporary postdoctoral positions, hoping to wait out the economic doldrums. Despite the unfavorable job market, most new physics Ph.D.s found science-related positions, although not necessarily long-term jobs in physics research. The dashing of expectations for traditional university research careers in the first half of the 1970s brought bitterness to some and changed the perspectives of many.† As the decade progressed, industrial employment grew in physics and in related science and engineering areas. Physics Ph.D.s adjusted their expectations and pursued the changing opportunities.

As new employment avenues began to expand in the 1970s, degree production in physics plummeted. Bachelor's degrees had already begun to decline by the late 1960s; Ph.D.s soon followed, dropping from over 1500 at the beginning of the 1970s to little more than 900 by the end of the decade. This reaction to the difficult market conditions of the early 1970s eased the employment situation for the new Ph.D.s emerging in the middle to late 1970s. Job offers increased; by 1980, only 2 percent of the Ph.D.s lacked job offers at the time their degrees were granted, compared with 25 percent at the start of the decade.

While this steep decline in physics degree production was taking place, the number of postsecondary-school students was rapidly growing. The children of the extended demographic baby boom of the 1950s, encouraged by the availability of educational loans, filled the colleges in the 1970s. Many students entered the biological and social sciences, but few majored in physics. Ph.D.s in physics, in fact, represented only 7 percent of all natural science and engineering doctorates in the late 1970s, down from the 11 percent figure that had been relatively constant for several decades. Although such a shift augured well for the competitive employment position of new physics-degree holders, it raised concern about the availability of future human resources in physics.

The one area where employment opportunities for new physics Ph.D.s had not altered by the beginning of the 1980s was academe. The young physicists who had swelled the ranks of assistant professors in the 1960s were now tenured, but they were far from retirement. In 1980, assistant professors represented only 14 percent of academic physics staffs, a lower percentage than in any other scientific discipline. With few senior positions opening, many of these assistant professors could not be awarded tenure. The effect of this "missing generation" of young academics is reflected in the dramatic aging of university physics department staffs from a median age of 38 in 1973 to 44 in 1981.

---

* See Supplement 3 on research funding for further details.

† *The Transition in Physics Doctoral Employment, 1960-1990* (American Physical Society, 1979).

ENROLLMENTS AND DEGREES: THE PROLONGED DECLINE

As noted, physics Ph.D. production declined rapidly throughout the 1970s, dropping to the low 900s by 1980, and hovering at that level since. In 1980 the first-year physics graduate student enrollment was 2439, down 44 percent from the middle 1960s. It appeared that graduate physics enrollments had finally bottomed out; 1981 saw the first rise in enrollments in 15 years. While it would be another 5 to 6 years before these increases could be reflected in Ph.D. production, an eventual turnaround seemed on the horizon.

U.S. and Foreign Composition

Detailed analysis of enrollments, however, pointed up a new phenomenon. In the 1970s foreign students constituted about one fifth of physics graduate students, or about 600 new foreign nationals a year. As the total number of first-year physics graduate students reached its nadir in 1980 and then began to rise, a major change in the citizenship composition of these students was also occurring. By 1983, more than 1000 of the first-year physics graduate students were foreign citizens—40 percent of that entering class. U.S. university physics departments were drawing students from many countries, providing particular opportunities for students from the less-developed countries in Asia and the Middle East. This influx of foreign students, however, clouded another trend. The decline in first-year graduate enrollment of U.S. citizens in physics did not, in fact, bottom out in 1980; it continued through 1983 (Figure S2.2).

If there had previously been concern about the availability of highly trained physics manpower in the United States, these data only heightened it. What lay behind the continued decline in U.S. graduate students? Although there were few academic opportunities, the general employment situation for new physics Ph.D.s was healthy. Job offers were plentiful, and starting salaries were high. Were the front-page breakthroughs in the biosciences, the new technology excitement of computer science, and the financial rewards of the professions drawing bright potential physics students away? Such changes in student career directions could have a major impact on a comparatively small area of concentration like physics.

In 1984, the number of first-year foreign graduate students continued to increase; however, U.S. enrollments for the first time in many years increased even more. Although 1985 saw a flattening of these trends, the changes in what had been a long decline in U.S. first-year enrollments provided at least a temporary sense of relief to the physics community.

Women and Minorities

Despite intensified efforts on the part of professional societies in physics and of related women's groups, the approximately 3 percent of Ph.D. physicists that women still represent (Figure S2.3) is the lowest among the major scientific areas. New physics Ph.D.s do include a higher proportion of women, but the actual number of doctorates awarded to women in the United States

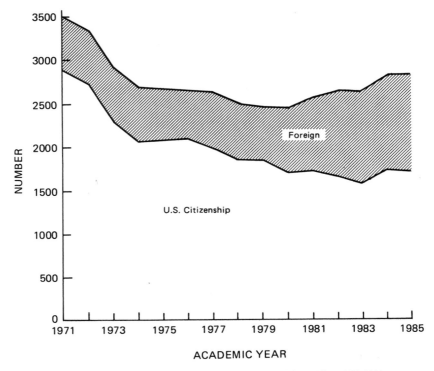

FIGURE S2.2 First-year graduate physics students by citizenship, 1971-1983.

has not changed since the middle 1970s—approximately 65 per year. In most science and engineering areas, increased degree production has reflected the growing participation of women. Enrollment of women in the formerly male bastions of engineering and computer science, particularly at the undergraduate level, has skyrocketed. The same cannot be said of physics.

The reasons for the continuing underrepresentation of women in physics are not yet fully understood. Women constitute the major untapped future resource for the physical sciences and engineering; nowhere is that more true than in physics. Whether one is thinking of the future development of the field or of equal access by all segments of the population, the minimal participation of women in physics remains a major concern.

If the representation of women in physics is low, the representation of indigenous U.S. minorities, e.g., blacks, Puerto Ricans, Mexican Americans, and Native American Indians, is even lower. They make up less than 1 percent of the physics labor force; total production at all degree levels remains low.

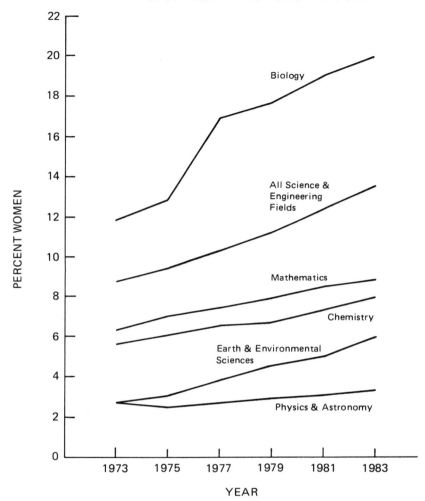

FIGURE S2.3 Women as a proportion of all doctoral scientists by selected fields, 1973-1983.

Full representation is found only among U.S. Orientals. This is a problem shared throughout the science and engineering community. Some inroads have been made at the bachelor's level in such fields as engineering and chemistry and, probably, computer science. Physics, with few such direct entry-level positions, has seen no increase in the representation of indigenous minorities. If anything, the number may be decreasing.

TABLE S2.1 Number of Physics Doctorates by Subfield, 1969-1983

Subfield	Academic Year														
	1969	1970	1971	1972	1973	1974	1975	1976	1977	1978	1979	1980	1981	1982	1983
Solid State	360	402	442	393	400	351	319	282	257	243	243	202	250	235	221
Elementary Particle	220	258	278	198	222	148	124	128	138	135	119	117	117	118	136
Nuclear[a]	188	212	227	234	182	144	130	96	93	77	103	73	62	53	90
Atomic and Molecular	127	152	124	150	122	120	138	116	105	88	72	69	65	96	71
Plasmas	61	85	86	93	74	57	53	75	72	68	62	59	65	69	72
Astrophysics	55	63	54	66	67	77	71	72	57	74	57	69	59	50	65
Optics	16	30	25	31	33	26	33	50	31	33	46	43	54	42	50
Acoustics	11	22	20	20	15	11	12	9	12	14	13	23	13	11	14
Fluids	24	21	20	27	28	22	22	20	14	13	14	15	14	13	15
Other	345	364	405	360	383	322	331	309	307	257	321	263	266	274	258
Total	1407	1609	1681	1572	1526	1278	1233	1157	1086	1002	1050	933	965	961	992

SOURCE: NAS.

[a] Between 1969 and 1982, this category was listed as Nuclear Structure. In 1983, it was changed to Nuclear Physics, which it was felt more accurately described the scope of the subfield. This change to a broader definition is probably responsible for the apparent upswing in degree production in this subfield between 1982 and 1983. The earlier dramatic decline in degree production in the late 1970s and early 1980s reflected only degrees awarded in nuclear structure, an underestimate of the total number produced in nuclear physics.

Declining Enrollments in Physics Subfields

The steep decline in physics Ph.D. production affected most of the major subfields of physics during the 1970s (Table S2.1). Degrees in solid-state physics; elementary-particle physics; nuclear physics; and atomic, molecular, and optical physics had dropped to less than half of their peak early 1970s levels by the beginning of the following decade. They have remained relatively stable since. Any continued erosion in subfield degree production may present threats to the research vitality of an area and could be a cause for concern in the entire physics community.

While degrees in plasma physics and astrophysics also declined from their peaks in the early 1970s, the drops were not so precipitous. Only in optics, one of the most applied of the major physics subfields, could a countertrend be observed; here, Ph.D. production doubled between 1971 and 1981. A similar increase was occurring in medical physics, although this area was not yet being monitored by the National Research Council.

RETENTION OF PHYSICS DEGREE HOLDERS—MOBILITY

Physics has traditionally trained practitioners for other disciplines: at the bachelor's level for engineering, at the master's level for education and business, and at the Ph.D. level for a multitude of developing science and engineering areas. More than any other scientific discipline, physics has sent its Ph.D.s to related disciplines to serve society's growing technological needs.

Much of the outward mobility of physics Ph.D.s has been into engineering and interdisciplinary areas such as geophysics, materials research, and biophysics; but Ph.D. physicists also work in areas ranging from chemistry to the biosciences. Some of this mobility occurred within academe where physicists teach and conduct basic research in related science and engineering departments. Most of it, however, occurred in the industrial sphere where applications of physics research move easily across disciplinary barriers.

While patterns of high outward mobility have been typical of physics and have actually alleviated potential problems during periods of surplus, one must re-examine their impact during periods of dwindling supply. In the early 1970s, 30 percent of Ph.D. physicists were working outside of physics. As the economic support for physics research declined and alternative opportunities arose, the proportion increased to approximately 40 percent. Since the middle 1970s, employment opportunities in physics have improved; nevertheless, physics continues to provide needed skilled resources to other science and engineering areas at the same high rate. As the 1980s began, a large number of Ph.D. physicists were moving into the growing areas of systems and electronic engineering and computer science. In 1981, 27,000 physics Ph.D.s were employed in the United States; 10,400 of them were working in nonphysics areas.

Physics will certainly continue to provide many of its well-trained researchers to the new, growing areas of the future. Concern, however, may arise when this high level of outward mobility is coupled with a decline in the supply of new physics degree recipients—the dominant pattern of the past decade.

AN AGING COMMUNITY

The declining physics Ph.D. production during the 1970s, coupled with the departure of many young Ph.D.s to neighboring science and engineering areas, has produced an aging of the physics community over the past decade. This aging was most marked in academe where opportunities for young physics faculty remained blocked by the highly tenured-in staff at most physics departments and the lack of funds for staff expansion (Figure S2.4).

In the early 1970s, physics faculties, at a median age of 38, may have been unusually young, reflecting the heavy influx of new Ph.D.s during the previous decade. This was certainly not true in 1981, by which time the academic physics community had aged by an average of 6 years to 44. They were not only the oldest group within the physics community but they were also older than academics in other science disciplines (Table S2.2). Most of the full professors who by 1981 made up the majority of the academic physics staffs were still more than a decade away from retirement. Thus, additional aging is expected in academe through the 1980s, although some expansion of opportunities appears likely as retirements slowly begin to increase.

The universities provide the major stimulus for the important basic research that drives the development of the entire field. An aging faculty raises a number of issues: the most obvious is the lack of academic opportunities for young

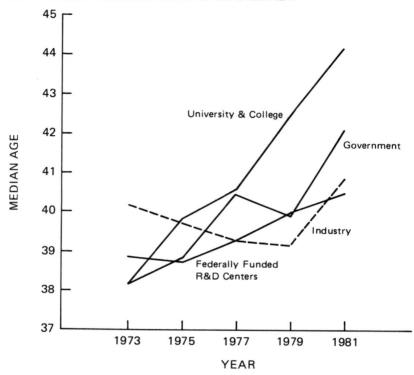

FIGURE S2.4 Changing median age of physicists by employment sector, 1973-1981.

physics Ph.D.s.* At the predoctoral, postdoctoral, and junior faculty levels, young physicists have traditionally made creative contributions to forefront basic research in physics.† The full effect of these contributions frequently does not appear for several decades; thus, many of the accolades of the 1980s have their source in the research activities of young physicists in the halcyon decade of the late 1950s and early 1960s. The relative absence of young researchers from today's academic scene is a cause for concern about our ability to pursue effectively the opportunities that lie ahead in physics.

Physicists have also aged in other areas of research. Except for the in-house government laboratories where funding cutbacks had been particularly large, however, such aging was minimal in contrast to that observed in the univer-

---

* *Physics Careers, Employment and Education*, AIP Conference Proceedings Number 39 (1978).

† See, for example, J. R. Cole and S. Cole, *Social Stratification in Science* (University of Chicago Press, Chicago, 1973), pp. 107-109; R. K. Merton, *The Sociology of Science* (University of Chicago Press, Chicago, 1973).

TABLE S2.2 Age Distribution of Full-Time Employed Ph.D. Faculty Members by Selected Departments, Spring 1980

Field	Age (Percentage)			
	Under 30	30-39	40-49	Over 50
All Fields	5	33	32	30
Physics	2	25	38	35
Chemistry	4	30	34	31
Engineering	6	28	36	29
Computer Science	12	43	26	18
Biological Science	3	35	29	33

SOURCE: National Science Foundation.

sities. In fact, the median age of the population of physicists employed in industry decreased during the 1970s as young physicists sought out new avenues of employment both within and outside physics.

CHANGING PATTERNS OF EMPLOYMENT

Patterns of physics employment have undergone major changes in the past decade. In the early 1970s, half of all physics Ph.D.s were academically employed. By 1981, the proportion had dropped to about 40 percent. Academic growth in physics during the decade was sluggish, and teaching staffs actually declined in size through 1979.

Employment of physics Ph.D.s in industry, on the other hand, burgeoned from under 5000 Ph.D.s in 1973 to more than 8500 in 1981, representing in the latter year nearly one third of all employed physics Ph.D.s. The steady growth in industrial employment of physicists reflected the favorable climate in that sector as well as the closing of academic doors. Although some of these new industrial opportunities were in physics itself, the areas of highest growth were in related sciences and engineering. The physicists who pursued these openings were, in general, primarily involved in development and design, and secondarily in applied research. Basic research has generally played a minor role in industrial employment. The proportion of industrially employed Ph.D.s engaged in basic research continued to decline over the period, falling to below 10 percent by the end of the 1970s.

Employment in the national laboratories, which had remained virtually unchanged during the otherwise expanding 1960s, increased steadily through the 1970s.* These laboratories offered special opportunities, in addition to

---

* National laboratories, such as Brookhaven and Fermi Laboratories, are completely funded by the federal government but are administered by universities, industry, and nonprofit groups.

providing a home for some of the basic physics research that had traditionally been carried out in the universities. However, the major increases occurred in the areas of applied research, as the national laboratories took on more diversified missions. In fact, by the beginning of the 1980s, a shift in emphasis at these laboratories from basic to applied research had occurred. Employment at in-house government laboratories showed only minimal growth throughout the decade.

These changing patterns of employment, particularly in the industrial sphere, allowed the absorption into the economy of the unusually large number of new physics Ph.D.s who made their appearance in the marketplace during the early 1970s. These Ph.D.s were not always doing what has been traditionally termed physics, but relatively few of them experienced any prolonged unemployment. In 1970, when *Physics in Perspective* was being developed, in the midst of a period of increasing economic hardship, the National Academy of Sciences report* noted:

> The 1970's have begun with recognition of a critical employment situation and the realization that the traditional patterns of employment in physics are changing. . . .If physicists understand the nature of this change, are flexible in their attitudes, and will undertake careers not only in physics but in related disciplines and specialties that also offer challenging problems, the outlook for the future could be more encouraging than was first assumed.

Physics Ph.D.s appear to have followed this dictum well. The entry of young physicists into the growing areas of applied research and technology provided an important stimulus for the development of a multitude of exciting new applications of physics research, from telecommunications through medicine.

Basic physics research did not enjoy the same healthy expansion, although a core of dedicated researchers did manage to weather the vagaries of an uncertain decade and continued to carve out new avenues of exploration at the forefront of physics. If basic research is to remain vital through the rest of the twentieth century, it is important to guard against any further erosion in this dwindled pool of human resources and to encourage the infusion of new talent.

PROJECTIONS

Projections of demand and supply are crucial to any understanding of future balances in scientific manpower. These projections are an attempt to portray what is most likely to happen to the physics labor force under different assumptions regarding expected changes in demand and supply. The projects should not be treated as predictions but rather as probable outcomes under given scenarios. Table S2.3 summarizes the critical parameters of the demand and supply models.

---

* Physics Survey Committee, *Physics in Perspective* (National Academy of Sciences, Washington, D.C.,1972), Vol. I, p. 856.

TABLE S2.3   Parameters in Demand and Supply Scenarios

Parameters in Demand Scenarios
  Demand for physicists in universities and colleges
    Openings, no growth
      Death and retirement: age structure
      Outmobility from academe
    Patterns of replacement
      Service course loads and demographics: future enrollments
      Promotion rates
      Age and tenure status of new hirees
  Demand for physicists in industry, government, and federal laboratories
    Openings due to replacement
      Death and retirement
      Outmobility from physics
    Openings due to growth
      Economic growth, R&D, and defense spending

Parameters in Supply Scenarios
  Physics Ph.D. production
    Demographic age factors and patterns of enrollment in higher education
    Perception of technological opportunities and choice of major
    Retention in the educational system
    Entry of foreign graduate students
  Available supply of new physics Ph.D.s
    Net migration of foreign nationals
    Immediate outmobility to other fields

Demand Projections

The demand for Ph.D. physicists in the future involves both replacements for current positions and growth, both directly in physics and indirectly in related disciplines. Because the expected demand scenarios in academe and outside it are quite different, they will be treated separately. Scenarios cover the period 1981-2001.* Estimated figures are presented for cumulative 5-year periods to minimize unpredictable year-by-year fluctuations.

ACADEME

The academic employment of Ph.D. physicists has undergone tumultuous changes during the past two decades. Few labor force analysts writing during

---

* The NAS biennial surveys have been used both to establish the employment patterns of physicists in the base year 1981 and to estimate mobility rates. 1981 is the last year for which data from the NAS biennial surveys are available.

the expansionary period of the 1960s envisaged the tight academic market that would dominate the physics scene throughout the 1970s and into the 1980s.* These abrupt changes have produced a unique age structure: in 1981, 40 percent of university physics faculty were over 50 years old, while another 40 percent were between the ages of 40 and 49. This middle-aged "bulge" has a major effect on any projections of future academic openings.

Death, retirement, outmobility, and promotion rates take on added importance when such large numbers of individuals are concentrated in the older age groups. We are assuming constant age-specific death rates for our academic professoriat, based on current TIAA-CREF mortality schedules for males.† However, while the age-specific death rates remain constant, the numbers of probable deaths and potential openings will escalate as the middle-aged bulge of 1981 ages further.

Retirement practices are not so clear-cut. The laws deferring mandatory retirement to age 70 were extended to academe in 1982. However, TIAA-CREF officials do not yet see any evidence of a further shift toward delayed retirement. If anything, they see some increased tendency for academics to seek early retirement. Thus our model allows for some early retirement and for half of the physics faculty members still working at age 65 to remain employed beyond that age. The combined effects of death and retirement should thus bring a dramatic increase in university employment opportunities in the 1990s.

Academic positions can also be freed by mobility. However, few tenured older faculty have left academe in recent years, and, conversely, not many senior-level physicists have been drawn into academe from industry or government. Movement into and out of academe by younger researchers, on the other hand, is an elastic and volatile phenomenon closely related to promotion rates and the availability of permanent senior positions. While American Association of University Professors rules set maximum time periods for which one can remain in untenured positions, they do not establish minimum times. If shortages develop, promotion policies are perhaps the most manipulable in the system. The model focuses on expected future openings in the senior ranks as the best indicator of long-term employment opportunities.

In addition to replacement needs, academic openings can result from growth. Total faculty demand is, we believe, more tied to service course loads than to basic research needs. By the mid-1990s, the 18- to 24-year-old age group, from which college students are traditionally drawn, will decline by over 20 percent. How many of them will attend college and take physics is, of course, the central issue. If tomorrow's students continue to need a strong technological background, then it is unlikely that enrollments in traditional

---

*Alan Cartter was less sanguine than many of his colleagues about the continuing expansionary needs of the future, but his voice was not the dominant one at the time.

† TIAA-CREF is the retirement annuity fund of the great majority of academics. Thus, their mortality schedules are most appropriate for use here. They will also be used when dealing with nonacademic physicists, since general mortality schedules would indicate death rates abnormally high for professional workers. In a community that is 95 percent male, the dependency on male death-rate schedules should not need comment.

science areas such as physics would suffer as precipitous a decline. However, it is questionable whether they would move so counter to the overall expected decline in student bodies as to produce a demand for faculty growth. Thus, the following scenarios assume that there will be no real growth in academic physics faculties through most of the remainder of the century.

DEMAND SCENARIOS—UNIVERSITIES

Our projections for university physics demand involve the following three major scenarios. In each scenario, replacement positions refer to existing senior positions vacated through death, retirement, and mobility out of academe.

- All replacement positions are filled.
- 10 percent of replacement positions become unavailable.
- 20 percent of replacement positions become unavailable.

All three scenarios recognize the expected decline in number of traditional university students in the late 1980s and 1990s and incorporate some countervailing swing toward increased demand for service courses in physics and other hard sciences. We believe that the second scenario, which assumes a moderate shift toward a more science-oriented student body, is the most likely. It should be recognized that any cuts in physics faculty will hit some institutions harder than others.

Under the three replacement scenarios, the median age of the university physics professoriat—44 in 1981—can be expected to continue to climb by an additional 4 to 6 years through the early 1990s, peaking at somewhat over 50. Thereafter, a rapid decline in median age is expected as staff from the middle-aged bulge of 1981 begin to retire and young academics fill out the professorial ranks.

Figure S2.5 illustrates the probable ranges in future median ages based on extrapolations from past hiring and promotion practices. By 2001, professors under 40 will increase to between one quarter and one third of the total physics staff, not so high a proportion as was found in the early 1970s but a more substantial base than is currently observed.

DEMAND SCENARIOS—4-YEAR COLLEGES

Physics faculties at the 4-year colleges are considerably smaller in number, and their age structure is notably younger, than are their counterparts in the universities. Possible openings due to death and retirement through 2001 are thus relatively limited. We expect the declining number of traditional college-age students enrolling by the 1990s to have a dramatic effect on the physics programs at the 4-year colleges, many of which are even now experiencing financial difficulties.

We present three possible scenarios, ranging from a severe staff reduction of 30 percent to a more modest one of 10 percent. Technological opportunities of the future and the relevance of a solid physics background may pull college students into physics courses at a higher rate than currently observed.

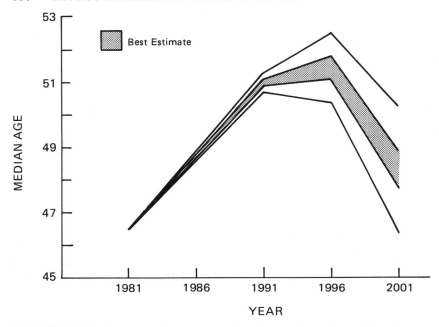

FIGURE S2.5 Projected median age of university physics professoriat at 5-year intervals, 1981-2001.

However, to achieve only minor staff reductions, physics departments at 4-year colleges will need to recruit more nontraditional students.

The accompanying tables for universities (Table S2.4) and 4-year colleges (Table S2.5) indicate the expected academic openings between 1981 and 2001 under the major scenarios outlined above. According to our intermediate scenario, tenured academic openings will increase steadily, swelling from 750 for the first 5-year interval (1981-1986), to nearly 1200 in 1996-2001. Whether new physicists will be available to fill these openings will be addressed later.

DEMAND SCENARIOS—INDUSTRIAL AND OTHER NONACADEMIC SECTORS

Opportunities for physicists over the coming decade will derive primarily from openings in industry, government laboratories, and nonprofit organizations, where 60 percent of Ph.D. physicists are currently employed. About half of these physics Ph.D.s are working directly in physics.

Since only 20 percent of these physicists were over the age of 50 in 1981, death and retirement will not be major sources of openings during the rest of the century. Most analysts, however, project some growth, in contrast to the projections for academe. Highest growth is expected among small consulting

TABLE S2.4 Projected Number of Physics Openings in Universities Resulting from Death and Retirement of Senior Staff, 1981-2001[a]

	Academic Years				
	1981-1986	1986-1991	1991-1996	1996-2001	Total 1981-2001
Number of Openings Due to Death and Retirement	769	911	1018	1119	3815
Number of Openings Filled by Replacement Scenario[b]					
Low	615	729	814	895	3053
Intermediate	692	820	916	1007	3436
High	769	911	1018	1119	3815

[a] Senior staff are defined as physics Ph.D.s employed as Associate or Full Professors in universities.

[b] The three scenarios: low, intermediate, and high reflect the hiring to fill 80, 90, and 100%, respectively, of senior staff openings.

TABLE S2.5 Projected Number of Physics Openings in 4-Year Colleges Resulting from Death and Retirement of Senior Staff, 1981-2001[a]

	Academic Years				
	1981-1986	1986-1991	1991-1996	1996-2001	Total 1981-2001
Number of Openings Due to Death and Retirement	114	154	243	299	810
Number of Openings Filled By Replacement Scenario[b]					
Low	39	53	84	103	279
Intermediate	64	87	137	168	456
High	89	120	190	234	633

[a] Senior staff are defined as physics Ph.D.s employed as Associate or Full Professors in 4-year colleges.

[b] The three scenarios: high, intermediate, and low reflect the reduction of the total physics professoriat in 4-year colleges by 10, 20, and 30%, respectively.

firms and medical laboratories.* Growth is also expected in the larger electronics and communications industries where many physicists are concentrated. While federal employment is not expected to be a major source of new positions, the national laboratories should continue to provide moderate growth in employment opportunities for physicists. We will use variations around the Bureau of Labor Statistics' (BLS) intermediate projections to describe expected growth in overall physics employment outside of academe.†

We pose two basic scenarios for growth in nonacademic physics employment. The first scenario assumes a moderately paced economy reflecting nonacademic growth at approximately the levels that occurred during the 1973-1981 period. This growth would average 3 percent per year, which is consistent with the BLS intermediate projection for physicists through 1995. The second scenario assumes a somewhat slower growing economy, reflecting a more conservative increase in nonacademic opportunities. Demand would be above replacement but below that suggested in the preceding scenario. This growth would average 2 percent per year. Based on expectations over the long term for an increasingly technological labor market calling on sophisticated research and development skills, we believe that the moderate growth scenario is the most likely forecast.

Positions opened by physicists leaving the field are also an important source of future employment. Currently the net outmobility of physicists is nearly 1.5 percent a year. If the supply of physics Ph.D.s falls short of the demand, the level of outmobility may decrease despite a strong continuing pull into other fields. Thus the model poses two levels of net outmobility: moderate outmobility consistent with current levels and low outmobility of approximately 1 percent a year.

Table S2.6 indicates the number of positions in physics expected to be open in the 1981-2001 period under these two basic scenarios. The potential demand for physics Ph.D.s in related areas of science and engineering is difficult to measure precisely because there is often only a fine line separating them from physics. Nevertheless, demand for physics Ph.D.s in these areas is also expected to increase in the next decade. Thus, there are several hundred potential nonphysics or interdisciplinary openings for Ph.D.s in addition to those described in the scenarios above.

Supply Projections

Projections of supply depend on a number of interdependent variables: the size of the age groups from which the supply is drawn, patterns of enrollments in higher education, choice of major, retention in the educational system, and career directions following receipt of the degree. Perceptions of economic and

---

* See, for example, *Industry—Occupational Employment Matrix*, BLS data tape, 1980, and *Problems of Small, High-Technology Firms*, NSF 81-305.
† *Occupational Projections and Training Data*, BLS, Spring 1984.

TABLE S2.6 Employment Opportunities for Ph.D. Physicists Working in Physics Outside of Academe by Demand Scenario and 5-Year Groupings, 1981-2001

	Academic Years			
	1981-1986	1986-1991	1991-1996	1996-2001
Conservative Growth (2% per year)				
Low Outmobility (1% per year)				
Replacement	803	939	1146	1362
New Openings	757	757	757	757
Total	1560	1696	1903	2119
Moderate Outmobility (1.45% per year)				
Replacement	963	1103	1305	1509
New Openings	757	757	757	757
Total	1720	1860	2062	2266
Moderate Growth (3% per year)				
Low Outmobility (1% per year)				
Replacement	803	960	1188	1428
New Openings	1136	1136	1136	1136
Total	1939	2096	2324	2564
Moderate Outmobility (1.45% per year)				
Replacement	963	1132	1364	1600
New Openings	1136	1136	1136	1136
Total	2099	2268	2500	2763

occupational demand also have an effect.* Long-term projections of possible changes are of particular import for physics, where the training pipeline frequently requires more than 10 years. Potential physics Ph.D.s of the middle 1990s are making their career decisions now. We will first present short-range projections for the production of Ph.D. physicists and then venture into the uncharted territory of the remainder of the twentieth century.

PHYSICS PH.D. PRODUCTION

Most of the new physics Ph.D.s of the late 1980s are already in the graduate school pipeline. Extrapolating from first-year graduate students at Ph.D.-

---

* See, for example, R. Freeman, *The Labor-Market for College-Trained Manpower* (Harvard University Press, Cambridge, Mass., 1971); R. Freeman, "Supply and Salary Adjustments to the Changing Science Manpower Market: Physics, 1948-1973," *American Economic Review* 65:27-39 (March 1975).

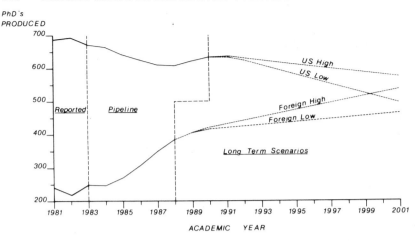

FIGURE S2.6 Physics Ph.D. production by citizenship, 1981-2001.

granting departments—a best past-production predictor—we expect 6800 physics Ph.D.s to be awarded between 1984 and 1990.* Note should be made of the projected increase in the foreign component, a reflection of the dramatic increase in first-year foreign graduate students since 1979 (Figure S2.6).

Projections of Ph.D. production through the remainder of the century involve analysis of the education pipeline leading to graduate school. The group of 18- to 24-year-old males, from which traditional undergraduate physics majors and subsequent bachelors are drawn, will have decreased by over 20 percent by 1996. However, because of the increasing emphasis on the utility of a scientific background in tomorrow's technological society, our probable scenario assumes a more modest decline in bachelor's degrees in physics of 10 percent from current levels. The assumption presupposes that steps will be taken to alleviate the currently deteriorating condition of precollege science education so that potential physics majors are not disenchanted with science in general before even entering college.

Approximately one third of recipients of bachelor's degrees in physics go on to physics graduate study. Another 15 percent of first-year physics graduate students come from other disciplines. Our model assumes that the latter relationships will remain relatively stable in the future.

---

* Physics Ph.D. students complete their programs in 4-9 calendar years at the following rates: 3, 7.5, 14, 11, 6, and 1.3 percent, respectively. Thus, on the average, approximately 43 percent of the first-year students eventually earn their Ph.D.s. The last 2 years of the short-term projections were supplemented by extrapolations of first-year graduate students from junior majors using linear regression analysis.

The participation of foreign graduate students, which has increased dramatically since 1979, will have an important influence on future Ph.D. production in physics. The scientific and technological needs of both developed and developing countries are expected to stay high during the remainder of the century. Our model thus assumes a growth in the number of foreign students of between 1 and 2½ percent a year through 2001. The effect on Ph.D. production will depend on the graduate-school completion rate of these students. In the final years of the century, we project between 980 and 1100 physics Ph.D.s a year (Figure S2.7). Under the most extreme scenario, U.S. citizens around the end of the century may become a minority among physics Ph.D. recipients.

SUPPLY OF PHYSICS PH.D.S: 1981-2001

Physics Ph.D. production does not equal supply. Two major factors that reduce this pool must be taken into account: immediate outmobility from physics and the return migration of foreign degree recipients. Many physics Ph.D.s—traditionally about 20 percent—move into nonphysics employment within 3 years of having received their degree. This outmobility increased to 30 percent during the difficult employment market of the early and middle 1970s, and it has not shown any marked decrease since. The current high level, however, seems to be guided more by pull than by push factors. Our model assumes that the outmobility will move toward the more typical 20 percent level as basic research opportunities in physics improve during the 1990s.

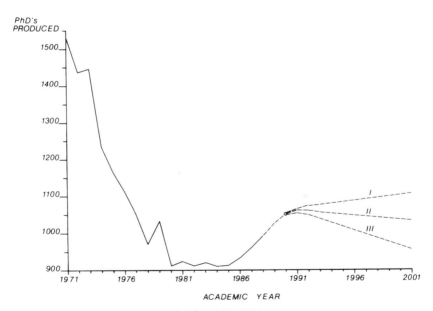

FIGURE S2.7   Physics Ph.D. production, 1971-2001.

TABLE S2.7 Physics Ph.D. Production and Probable Supply by 5-Year Groupings, 1981-2001

Academic Years	Total Number of New Ph.D.s	Total Number of Foreign Ph.D.s	Estimated Number Available to Labor Force[a]
Reported and Pipeline Projections			
1981-1986	4591	1296	2760-3150
1986-1991	5096	2985	2870-3280
Long-Term Scenarios[b]			
I. Low U.S.-Low Foreign			
1991-1996	5140	2170	2840-3250
1996-2001	5190	2580	2730-3120
II. Low U.S.-High Foreign			
1991-1996	5260	2170	2890-3310
1996-2001	5190	2580	2730-3120
III. High U.S.-Low Foreign			
1991-1996	5260	2170	2920-3340
1996-2001	5200	2280	2840-3250
IV. High U.S.-High Foreign			
1991-1996	5400	2320	2970-3400
1996-2001	5500	2580	2950-3370

[a] This column indicates the supply of Ph.D.s available after return migration of foreign citizens (assumed to be 50%) and immediate outmobility from physics of new Ph.D. recipients (assumed to range between 20 and 30%).
[b] Low-High U.S.: 1st-year physics graduate students who are U.S. citizens are assumed to decline 22% and 10%, respectively, between 1985 and 1997. Low-High Foreign: 1st-year physics graduate students who are foreign citizens are assumed to increase by 1% and 2.5% per year, respectively, over current levels between 1985 and 1997.

EDUCATION AND SUPPLY OF PHYSICISTS 113

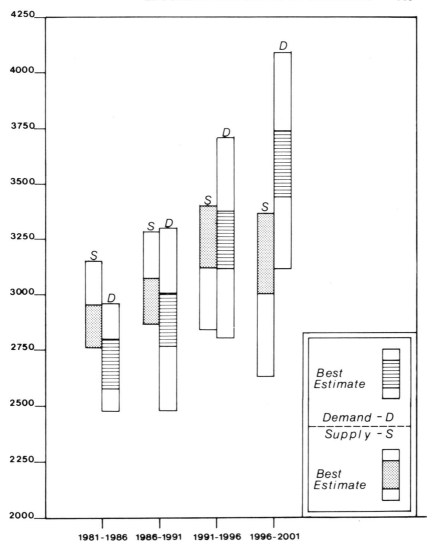

FIGURE S2.8 Projected demand for and supply of physicists by 5-year period, academic years 1981-2001.

The second factor affecting the supply of new physicists is the return home of new Ph.D.s who are foreign citizens. We estimate on the basis of past experience that one half of these foreign nationals will return home, either immediately after receiving their degrees or following an ensuing postdoctoral or other temporary position. The phenomenon is sensitive to change. With the foreign component potentially approaching half of the Ph.D. production, such changes could have a major effect on the available supply. While the number of U.S. citizen physics Ph.D.s is expected to decline, the supply of foreign Ph.D.s could double by 2001 (Table S2.7).

The Demand-Supply Balance

Figure S2.8 illustrates the projected range of supply of new physicists and demand for Ph.D.s to work in physics. As discussed, forecasting either supply or demand involves many parameters, some unknowable. The shaded portions of the figure reflect our best estimates of the likely demand and supply levels during each of the four 5-year periods through the remainder of the century. We project a balance between likely demand and supply through 1991. Later in the 1990s, however, it appears that the supply of new physicists may not match the number of possible physics openings. If this mismatch occurs, there is likely to be increased competition among employers for well-trained physicists.

The divergence between probable demand and supply may be even greater than we have indicated because it does not include the projected increase in employment opportunities in the many neighboring scientific and engineering areas where physicists have always made contributions of major import.

CONCLUSION

If the assumptions outlined in the previous pages hold over the remainder of the twentieth century, we project a precarious balance between demand and supply through the early 1990s. Beyond that, there are likely to be more opportunities than there are new physicists to fill them. There will be keen competition among disciplines for the brightest students and among employers for new Ph.D.s. Under each of our scenarios, employers of physicists are seen as becoming increasingly dependent on non-U.S. citizens. Given the large numbers, any changes in the retention of the foreign component could have a strong effect on the overall demand and supply picture.

The rest of the century represents a broad span of time during which the activities of the physics community, the federal government, and industry could certainly help to alleviate any pending shortages. The excitement now inherent in physics and the current opportunities for further research are great. It would be unfortunate if an insufficient supply of qualified physicists left these opportunities less then fully realized. Now is the time for the physics community to seriously consider means for attracting the bright students of tomorrow. As we do so, we should recognize that the largest untapped pool of potential physicists will be found among those groups underrepresented in the past.

Supplement 3

Organization and Support of Physics

In the United States, basic physics is supported by many organizations and a variety of funding patterns. These patterns reflect such factors as industrial interest, relevance to agency missions, and degrees of centralization, which can range from intense concentration in large national facilities to highly dispersed small research groups.

In this supplement, we describe the funding patterns, their differences and similarities, and some of their consequences. We point out the increasing need for centralized facilities in some areas, the increasing need for dispersed support in other areas, and certain trends in physics funding during the past 15 years.

This supplement focuses on the organization and support of physics as a whole. Specific problems of the various subfields are highlighted, but the reader should consult the accompanying panel reports for a thorough discussion of any particular one.

THE DIVERSITY OF INSTITUTIONS FOR RESEARCH IN PHYSICS

At present, more than half (53 percent) of Ph.D. physicists work in academic institutions; the remainder are in industry (21 percent) and in government and national laboratories (26 percent). An increasing number of industrial and governmental physicists use specialized facilities at regional or national laboratories for part of their research, but most by far of the physical research in the United States is carried out by small groups, often in universities. As emphasized in Chapter 3, it is important to recognize that small-group research represents one of this nation's major strengths in physics. It plays a major role in the advance of physics, and the research contributes directly to our national programs and the generation of new technology.

Major Facilities and National Laboratories

It is probably fair to say that big science—the collaborative use of large centralized facilities at national laboratories—had its origins in the Manhattan Project. However, even without this wartime stimulus the evolution of physics would have been, as it is, toward increasingly complex problems whose solutions frequently require larger, more sophisticated, and more expensive facilities. Elementary-particle physics has moved farthest in this direction: its research is concentrated in no more than four accelerator facilities in the United States. Nuclear physics, while clearly moving in the same direction, has so far managed to justify the maintenance of a broader base of facilities. Even in subfields such as atomic physics and condensed-matter physics, which pride themselves on their independent, self-sufficient, and individually sized research groups, some of the research now requires user programs with such facilities as synchrotron light sources, high-voltage electron microscopes, reactors, intense-pulse neutron sources, and the National Science Foundation's (NSF's) Materials Research Laboratories.

A very different example of the trend toward centralized facilities has been the development of national computing centers and associated networks. The NSF's National Center for Atmospheric Research (NCAR) and the Department of Energy's Magnetic Fusion Energy Computer Center (MFECC) make available to an individual researcher far more computer capability than could be justified or afforded on a more local level. Recently, the NSF has conducted a number of studies of the problem of scientific computing in general and large-scale scientific computing in particular (*Panel on Large Scale Computing in Science and Engineering*, P. Lax, Chairman, December 1982; and *Working Group on Computers in Research*, K. Curtis, Chairman, July 1983). The NSF has created four centers for computation in order to make large computers available to the scientific community.

The national laboratories play a vital role in advancing physics in the United States. They carry forward basic research missions and fill a variety of special needs. Often they can provide necessary facilities and services in cases where the size or cost is beyond the scale possible for individual research groups. They can also provide facilities that are too hazardous or too specialized for local laboratories. In developing, maintaining, and making these facilities available to outside researchers, the national laboratories play an important stewardship role. In order for these laboratories to maintain the quality of expertise necessary to perform this role, it is important for their staffs to be permitted and encouraged to spend a reasonable fraction of their time pursuing their own independent research.

In addition to providing the essential tools for many areas of basic research, the national laboratories carry out diverse missions of programmatic research in areas such as metrology, environmental monitoring, and calibration and standardization. Because of continuing changes in these missions and their priorities, there has recently been discussion* of the appropriate roles of these

---

* See, for example, *The Department of Energy Multiprogram Laboratories*, E.R.A.B. Report DOE/S-0015 (September 1982), and *Report of the Federal Laboratory Review Panel* (D. Packard, Chairman), White House Science Council (May 1983).

laboratories and of the best way for the federal government to manage these resources, while maintaining the flexibility to respond to changing research needs and to exploit new scientific and technical developments.

The Packard report makes a number of recommendations for increasing the effectiveness of the laboratories. The recommendations are particularly concerned with giving the laboratories more financial stability and flexibility. It is suggested that a substantial part (5 to 10 percent) of the laboratories' budgets be earmarked as discretionary funds for the director to use in exploiting innovative scientific opportunities. While seeking in this way to reduce the level of micromanagement by the funding agencies, the recommendations would at the same time make directors more accountable for the quality, productivity, and relevance of their laboratories.

The Packard report is also concerned with the establishment of clearly defined missions for each laboratory (beyond simple self-preservation) and with the strengthening of the interaction of federal laboratories with their users and with both industrial and university scientists and laboratories. Maintaining strong national laboratories is vital to the future of physics research in the United States.

University Research

In addition to their role in educating and training graduate students who will be the next generation of research scientists, university faculty and their laboratories carry out much of the basic physics research in the United States. Free of the responsibilities for programmatic research, the encumbrances of commercial justification, and, generally, restrictive commitments to large in-house facilities, the university component of physics research *can* be imaginative and flexible in pursuing new ideas. Because of this flexibility, and because of the diversity resulting from the large number of independent university researchers with interests covering the full range of physics, university laboratories have been able to compete successfully with national and industrial laboratories in spite of the overwhelming advantage that the other two components frequently have in terms of instrumentation and facilities.

Industrial Research

Industrial research in the United States is concentrated in condensed-matter physics; in atomic, molecular, and optical physics; and in interdisciplinary areas such as materials science and biophysics. In condensed-matter physics, industrial research constitutes nearly one third of the total effort in the United States. Industrial interest in these fields is due to the close relationship between basic discoveries in these areas and commercial application in electronics and optics. Industrial laboratories in the United States are responsible for much basic research, and they also provide a bridge between the research community and the development of new technologies. Such advances have allowed the United States to lead the world in the area of high technology.

Industrial laboratories like those at Bell and IBM are unique in the world, and they represent a national resource that could not easily be replicated. In interface areas, industrial research laboratories can provide unique opportunities and instrumentation for basic research. Industrial laboratories can frequently apply a broader range of techniques to a particular problem than is possible in universities. This can be crucial in materials work, where several characterization measurements are important in uncovering the physics of a specific problem.

THE COMPLEMENTARY ROLES OF OUR RESEARCH INSTITUTIONS

Industrial laboratories, national laboratories, and university laboratories have their particular strengths, which *complement* each other and should be used to ensure the vitality of our national research effort. As discussed further in the last section of this supplement, an important strength of U.S. physics research is the flexibility that results from the collaborative interaction and constructive cross-fertilization among the university, industrial, and national laboratory sectors.

A constructive, complementary relationship also exists between the large national facilities and the independent individual research groups. As discussed in Chapter 3, *balances* must be achieved and maintained between the needs of large national facilities and those of individual group researchers and between a several-billion-dollar, 20-40 TeV Superconducting Super Collider to explore the nature of matter and energy and a $200,000, ultrahigh-vacuum system for hyperthermal (1-1000 eV) ion-beam studies of the atomic structure of metal surfaces. One difficulty in maintaining such balances is the different visibility inherent in the different components and the absence of opportunities for launching major research initiatives in small-group research. A new, larger accelerator or a major new telescope can have an invigorating effect on an entire field, which extends far beyond the immediate users. It is difficult, if not impossible, to launch such an initiative when the goals of a field are diverse and the major need is not for a central facility but for increased support of small groups.

The growing role of research in the user mode at large national facilities can cause tensions. The situation was summarized in the 1970 report of the Physics Survey Committee,* and the description is just as valid today:

... As experimental work shifts from smaller local facilities to larger ones of national or international significance, it is clear that the pressures for the greatest possible efficiency in the use of the facility mount rapidly, as does competition for access to it. Under these circumstances it is difficult to allow a student a major role in the overall design of an experiment, its execution, and its analysis. It also is difficult, when scheduling is tight and long-term, to offer sufficient time and flexibility to allow a student to follow his curiosity into whatever new channels unfold during the course of an experiment. What is involved here is a different style of experimental educational experience and one that is intrinsically more specialized than has been traditional in physics. But there is little or no choice; physics departments must accept such changes if they are to remain active at the research frontiers that attract some of the most able students.

---

* *Physics in Perspective* (National Academy of Sciences, Washington, D.C., 1972), Vol. I, p. 599.

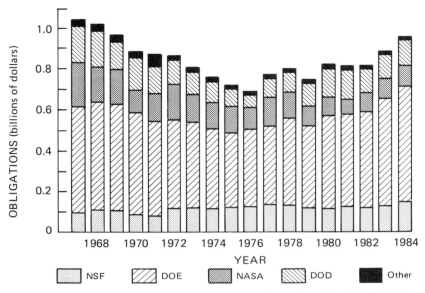

FIGURE S3.1 Federal obligations for basic research in physics, FY 1967-FY 1984.

FUNDING SUPPORT FOR PHYSICS RESEARCH

As shown in Figure S3.1 and Table S3.1, during the 14 years since the last Physics Survey, federal funding for basic research in physics (when expressed in *constant* dollars*) went through an initial decline from which it has gradually

---

* In order to provide what we hope is a consistent set of raw data and to make the connection with historical records more convenient, the tables in this supplement list the various detailed funding data in "current" dollars (referenced to the fiscal year for which they were appropriated). For the purposes of making meaningful comparisons between the data for different years, wherever reasonable and possible throughout this supplement, we have tabulated and graphed aggregate financial support figures referenced to FY 1983 dollars. To make this conversion, we have made use of the CPI-W index (urban and clerical workers); this decision was made partly on the basis that the largest fraction of this support is directly or indirectly related to these salaries and partly on the basis of consistency since this index has been widely used in previous studies. A much more detailed index has been developed (*High Energy Physics, Analysis of Cost and Price Changes*, HEP-81402, Division of High Energy Physics, DOE), which tracks separately the inflation of salaries, power, scientific equipment, construction, etc. A comparison of these two indices (CPI-W versus HEP) shows that—although the weighted average of the HEP (Operating/Capital Equipment/Construction) indices may differ from the CPI-W index by as much as ±2½ percent in any given year—when averaged over the period 1976-1981 (for which complete comparison data exist), the two indices agree to within less than 0.1 percent.

TABLE S3.1 Federal Obligations for Basic Research in Physics, Fiscal Years 1967-1984 (in Millions of Dollars)

Fiscal Year	1967	1968	1969	1970	1971	1972	1973	1974	1975	1976	1977	1978	1979	1980	1981	1982	1983	1984
							Current Dollars											
NSF	31.7	37.8	37.9	32.9	32.6	49.0	52.7	55.0	65.4	71.1	83.2	86.4	85.6	95.6	113.6	117.8	131.0	156.5
DOE	173.6	182.3	189.2	192.0	186.7	181.9	181.9	185.6	191.2	213.6	232.1	277.5	287.7	369.2	409.7	454.8	524.4	582.3
NASA	72.4	59.6	60.9	41.7	54.5	71.6	58.5	60.8	67.9	60.2	85.2	81.3	69.6	73.3	64.9	89.2	96.2	104.9
DOD	60.0	61.2	49.9	60.7	52.8	49.5	47.5	47.8	44.9	34.0	54.7	62.1	79.1	111.2	128.4	111.0	118.2	130.2
Other	11.4	12.0	13.0	11.7	24.1	9.8	10.1	11.1	9.5	9.4	12.2	11.5	13.7	18.9	18.8	17.9	18.5	17.7
Total	349.0	352.9	350.9	338.9	350.7	361.7	350.8	360.3	379.0	388.4	467.4	518.8	535.6	668.2	735.4	790.7	888.3	991.6
						Deflated 1983 Dollars (Using CPI-W)												
NSF Deflated	94.8	109.5	104.8	85.9	81.0	117.3	121.3	116.3	124.5	126.4	137.4	133.5	119.7	117.7	126.0	121.8	131.0	151.6
DOE Deflated	519.8	528.3	523.1	501.2	463.3	435.5	419.1	392.3	363.9	379.5	383.5	428.5	402.3	454.7	454.4	470.4	524.4	563.9
NASA Deflated	216.8	172.6	168.4	108.8	135.2	171.5	134.8	128.4	129.1	107.0	140.8	125.5	97.3	90.3	72.0	92.3	96.2	101.6
DOD Deflated	179.6	177.3	138.0	158.4	131.0	118.5	109.5	101.0	85.4	60.4	90.4	96.0	110.5	136.9	142.3	114.8	118.2	126.1
Other Deflated	34.1	34.8	35.9	30.7	59.8	23.6	23.3	23.5	18.2	16.8	20.2	17.8	19.1	23.2	20.8	18.5	18.5	17.1
Total Deflated	1045.2	1022.6	970.2	884.9	870.3	866.4	808.0	761.6	721.1	690.2	772.4	801.3	748.9	822.8	815.5	817.9	888.3	960.4

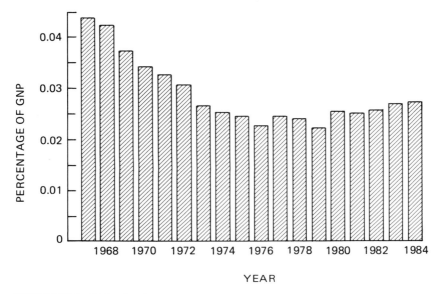

FIGURE S3.2 Federal obligations for basic research in physics (as percentage of GNP, FY 1967-FY 1984).

recovered in recent years. This pattern, and particularly the recovery, has not been uniform among the subfields. Furthermore, it should be realized that the CPI-W price index, like all such indices, does not adequately account for inflation, because of the need for more sophisticated and expensive equipment. Thus a constant level of funding over a decade can actually portray a deteriorating research condition.

While the overall federal support for basic physics research appears to have largely recovered to its level in the late 1960s and early 1970s when expressed in FY 1983 dollars, this support has not kept pace with the increasing GNP (Figure S3.2). Since 1967, federal support for basic physics research has fallen from nearly 0.044 percent of the GNP (1967) to a current level of 0.027 percent (1984).

Federal support for *applied* physics research (Figure S3.3 and Table S3.2) shows an initial decline after 1967, followed by a very striking real increase to a level that in FY 1984 is nearly twice what it was in FY 1970 (when expressed in *constant* dollars). It should be noted, however, that this real increase in support for applied physics is not broad; it is almost all due to the increase in the DOE fusion program beginning in FY 1976 and to more recent increases at DOD. Without these specific projects, the overall trend in Figure S3.3 would look similar to that portrayed in Figure S3.1. It should also be noted that these two sets of data (basic research and applied research) are aggregates reported as such by the individual agencies; they are not summations of the subfield data

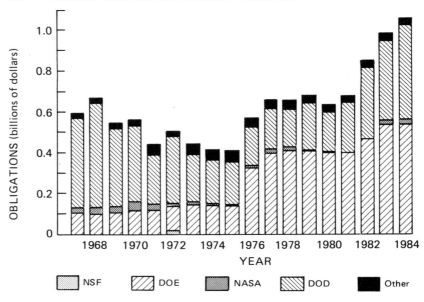

FIGURE S3.3 Federal obligations for applied research in physics, FY 1967-FY 1984.

presented below. The two different sets of data (basic and applied versus subfields) are each *internally* consistent; inconsistencies between the two sets are largely due to questions of definition. With this disclaimer in mind, we can use Figures S3.1 and S3.3 and their respective tables (Tables S3.1 and S3.2) to characterize the trends in the overall support for basic and applied physics research.

As discussed earlier, research is also performed in industrial laboratories and supported by private industry, particularly in the areas of condensed-matter physics and of atomic, molecular, and optical physics. Because of the problem of distinguishing between basic and applied research and development, financial support in this sector is generally not so easy to identify and define. The expenditures plotted in Figure S3.4 and listed in Table S3.3 were compiled by the NSF Division of Resource Studies. While the absolute level of this support may be open to question because of problems associated with definition, the level represented by these NSF statistics is quite consistent with the facts, noted earlier in this supplement, that 21 percent of the working physicists are in industry and that approximately 26 percent of the papers by U.S. authors published in *Physical Review Letters* and *Applied Physics Letters* are written by industrial physicists. In any case, these statistics should be internally consistent. They indicate that the 30 percent decline over the period 1970-1981 is real. This is a particular cause for concern in areas where the support is concentrated, such as in condensed-matter physics and atomic, molecular, and optical physics, because industry may in fact represent as much as one third of the total basic research.

TABLE S3.2 Federal Obligations for Applied Research in Physics, Fiscal Years 1967-1984

	\multicolumn{18}{c}{Fiscal Year}
	1967	1968	1969	1970	1971	1972	1973	1974	1975	1976	1977	1978	1979	1980	1981	1982	1983	1984
	\multicolumn{18}{c}{Current Dollars}
NSF	0.0	0.0	0.0	0.2	0.0	8.3	0.3	0.0	0.0	2.1	1.7	2.4	3.4	1.8	0.9	3.2	3.8	3.9
DOE	35.1	34.3	38.3	44.1	47.7	48.9	62.2	66.6	73.6	180.7	237.2	262.2	286.9	321.3	357.2	447.7	533.3	552.2
NASA	8.6	11.4	11.5	17.2	12.6	5.9	6.0	5.4	3.9	6.7	13.8	13.1	3.7	3.9	2.2	2.0	21.0	24.9
DOD	147.1	176.3	138.1	142.5	96.5	137.8	101.0	99.8	109.0	107.6	119.8	118.2	165.5	158.2	220.5	335.0	387.2	472.7
Other	7.5	9.0	10.2	11.6	21.3	10.9	23.0	24.6	29.6	23.8	26.4	27.8	27.7	29.2	29.6	31.8	33.4	32.9
Total	198.3	231.0	198.1	215.6	178.1	211.8	192.5	196.3	216.0	320.8	398.8	423.6	487.1	514.4	610.2	819.8	978.7	1086.7
	\multicolumn{18}{c}{Deflated 1983 Dollars (Using CPI-W)}
NSF Deflated	0.0	0.0	0.0	0.6	0.0	19.9	0.6	0.0	0.0	3.7	2.8	3.6	4.7	2.2	0.9	3.3	3.8	3.8
DOE Deflated	105.2	99.5	105.8	115.2	118.3	117.2	143.2	140.8	139.9	321.0	391.9	405.0	401.2	395.6	396.1	463.1	533.3	534.8
NASA Deflated	25.8	33.0	31.8	45.0	31.1	14.1	13.9	11.3	7.4	11.9	22.9	20.2	5.2	4.8	2.4	2.1	21.0	24.2
DOD Deflated	440.5	510.9	381.8	372.1	239.5	330.1	232.6	211.0	207.3	191.2	197.9	182.5	231.4	194.8	244.5	346.5	387.2	457.9
Other Deflated	22.3	26.2	28.2	30.2	52.9	26.1	53.1	51.9	56.3	42.3	43.6	43.0	38.7	36.0	32.8	32.9	33.4	31.9
Total Deflated	593.8	669.5	547.6	563.0	441.8	507.4	443.3	415.0	410.9	570.0	659.1	654.3	681.1	633.4	676.7	847.9	978.7	1052.4

SOURCE: NSF Division of Science Resource Studies.

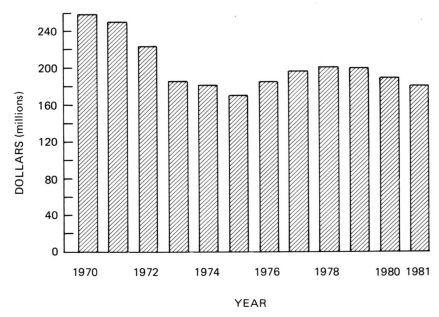

FIGURE S3.4 Industrial expenditures on basic research in physics, 1970-1981.

Two subfields for which complete, detailed data are available covering this entire period (FY 1970-FY 1984) are elementary-particle physics and nuclear physics; the data for these subfields are presented in Figure S3.5 and Table S3.4, and Figure S3.6 and Table S3.5, respectively. These data include the operating and equipment funds for both the experimental and theoretical components of the fields. Because of the different ways in which the agencies organize and report their research support, it was necessary to work with the program officers to separate out the different components and combine them in consistent ways. NSF normally reports Theory as a separate aggregate undifferentiated by subfield, and so we had to develop a consistent way of apportioning this. Although it would be desirable to present investments in

TABLE S3.3 Industrial Expenditures for Basic Research in Physics, 1971-1981 (in Millions of Dollars)

	71	71	72	73	74	75	76	77	78	79	81	81
Physics	111	112	94	83	91	92	116	121	(133)[a]	146	(155)	165
Physics Deflated	258	251	223	186	181	171	185	197	(212)	211	(189)	181

SOURCE: NSF Division of Science Resource Studies.
[a] Parentheses indicate estimated amounts.

ORGANIZATION AND SUPPORT OF PHYSICS   125

capital equipment separately from operating expense, neither NSF nor DOE handle Equipment consistently as a separable item. DOE has an equipment category, but this is used primarily for its national laboratories; the purchase of similar items by its contract universities is included inextricably in their operating funds, as in the case of NSF grants. Therefore, in order to be consistent in our tables and graphs, we have included all equipment funds together with operating funds. Although NSF likewise does not differentiate its Construction funds in its reports, these funds are usually more easily identifiable, and we have been able to separate them from the operating funds. Construction funds are discussed separately later in this supplement.

The elementary-particle physics data, Figure S3.5, reveal a more uniform level of support for operating funds than is shown for all of physics in Figure S3.1, although there were large fluctuations in the equipment fund. For nuclear physics, the plot of funding data in Figure S3.6 also shows a relatively more uniform pattern of support over the 15-year period than Figure S3.1 shows for physics as a whole.

The relatively constant funding levels ~$370 million (±10 percent) for elementary-particle physics and ~$157 million (±7 percent) for nuclear physics during this period should not be taken to mean that the facilities and operations have been static. In fact, in both of these subfields, there have been major facility closings and realignments of funding support. Of the seven elementary-particle accelerators operating in the United States in 1970, four (PPA, CEA,

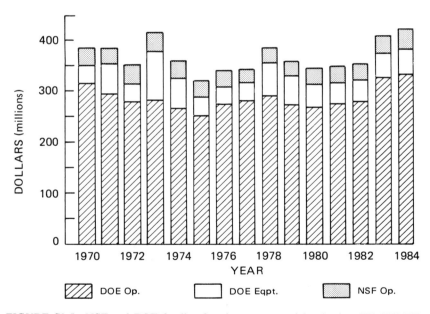

FIGURE S3.5 NSF and DOE funding for elementary-particle physics, FY 1970-FY 1984. Funding is expressed in FY 1983 dollars using CPI-W inflation factor. Construction funds are not included.

TABLE S3.4 NSF and DOE Funding for Elementary-Particle Physics, Fiscal Years 1970-1984, Compensated for Inflation Using CPI-W (in Millions of Dollars)

	Fiscal Year														
	1970	1971	1972	1973	1974	1975	1976	1977	1978	1979	1980	1981	1982	1983	1984
Operating and Capital Equipment															
DOE HEP Operating Budget	113.5	111.5	109.4	115.4	118.4	124.5	146.4	161.4	178.2	184.8	207.2	235.0	255.6	308.6	324.6
DOE Theory	7.0	7.0	7.0	7.2	7.4	7.7	8.0	8.6	9.5	9.7	10.7	12.4	13.6	17.2	17.9
DOE Total Operating	120.5	118.5	116.4	122.6	125.8	132.2	154.4	170.0	187.7	194.5	217.9	247.4	269.2	325.8	342.5
DOE Total Operating Deflated*	314.6	294.1	278.8	282.4	265.9	251.5	274.3	280.9	289.9	272.0	268.3	274.3	278.4	325.8	331.7
DOE Equipment	13.7	23.9	14.5	41.7	28.0	19.1	18.8	21.8	42.0	41.2	36.0	37.5	40.7	47.5	51.5
DOE Equipment Deflated*	35.8	59.3	34.7	96.1	59.2	36.3	33.4	36.0	64.9	57.6	44.3	41.6	42.1	47.5	49.9
NSF Operating	11.2	11.6	13.7	14.0	14.0	15.0	16.1	13.2	16.2	16.8	22.6	25.2	26.2	28.7	35.9
NSF Theory	2.4	1.4	2.4	2.4	2.2	2.4	2.4	2.9	3.6	3.8	3.7	4.5	4.7	5.2	6.0
NSF Total Operating	13.6	13.0	16.1	16.4	16.2	17.4	18.5	16.1	19.8	20.6	26.3	29.7	30.9	33.9	41.9
NSF Total Operating Deflated*	35.5	32.3	38.6	37.8	34.2	33.1	32.9	26.6	30.6	28.8	32.4	32.9	32.0	33.9	40.6
Grand Total[a]	147.8	155.4	147.0	180.7	170.0	168.7	191.7	207.9	249.5	256.3	280.2	314.6	340.8	407.2	435.9
Grand Total Deflated	364.0	366.4	329.6	394.9	341.3	305.3	326.2	333.0	374.6	350.2	339.0	345.6	351.4	407.2	422.2
Construction															
DOE Construction	81.0	70.0	47.0	45.0	15.0	2.0	8.0	32.0	50.0	62.0	66.0	65.0	55.0	58.0	81.0
NSF Construction	0.0	0.0	0.0	0.0	0.0	0.0	0.0	6.0	7.6	6.0	0.0	0.0	0.0	0.0	0.0
Total	81.0	70.0	47.0	45.0	15.0	2.0	8.0	38.0	57.6	68.0	66.0	65.0	55.0	58.0	81.0
Total Deflated	211.5	173.7	112.6	103.7	31.7	3.8	14.2	62.8	89.0	95.1	81.3	72.1	56.9	58.0	78.4

NOTE: Graph is constructed from rows with asterisks.
[a] Grand total includes operating, theory, and equipment.

ORGANIZATION AND SUPPORT OF PHYSICS 127

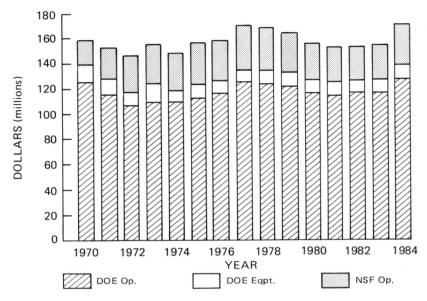

FIGURE S3.6 NSF and DOE funding for nuclear physics, FY 1970-FY 1984. Funding is expressed in FY 1983 dollars using CPI-W inflation factor. Construction funds are not included.

ZGS, and Bevatron) were retired from research in elementary-particle physics, while one new one (FNAL) began active research. This centralization of elementary-particle physics facilities is expected to continue. In nuclear physics, there was a reduction in the number of active accelerator facilities, from 89 in 1969-1970 to 27 in 1984; of these 27 remaining nuclear-physics facilities, 14 have had or are now undergoing major upgrades or have installed completely new accelerators. The current proposals for constructing two new large nuclear-physics facilities (CEBAF and RNC) indicate that the trend toward centralization will undoubtedly continue.

The funding trends for the six subfields are plotted in Figure S3.7 (Table S3.6) for the past 5 years (1980-1984); this is the only period for which we were able to extract consistent data for all the subfields. While it is clear from this plot that over the 5-year period there have been real increases in each of these subfields, it should be remembered that in general these do *not* represent the funding patterns over the entire 15-year period (1970-1984) since the last Physics Survey. The real growth shown in Figure S3.7 is in general only a partial recovery from the funding reductions experienced earlier in this period (Figures S3.1, S3.5, and S3.6). The real growth in the past 5 years ranges from +9 percent for nuclear physics to +42 percent for cosmic and gravitational physics and +43 percent for plasma and fluid physics.

Cosmic and gravitational physics is an emerging subfield that has just begun to develop facilities for gravitational radiation detection and satellite experi-

TABLE S3.5 NSF and DOE Funding for Nuclear Physics, Fiscal Years 1970-1984, Compensated for Inflation Using CPI-W (in Millions of Dollars)

	Fiscal Year														
	1970	1971	1972	1973	1974	1975	1976	1977	1978	1979	1980	1981	1982	1983	1984
Operating and Capital Equipment															
DOE HEP Operating Budget	45.4	43.8	41.7	44.6	48.7	55.7	61.4	70.9	74.6	80.9	88.2	95.7	104.7	107.9	122.0
DOE Theory	2.6	2.7	2.8	3.0	3.1	3.4	4.1	4.9	5.4	6.1	6.1	7.0	7.7	8.2	9.0
DOE Total Operating	48.0	46.5	44.5	47.6	51.8	59.1	65.5	75.8	80.0	87.0	94.3	102.7	112.4	116.1	131.0
DOE Total Operating Deflated*	125.3	115.4	106.6	109.6	109.5	112.4	116.4	125.3	123.6	121.6	116.1	113.9	116.3	116.1	126.9
DOE Equipment	5.4	5.2	4.5	6.4	4.3	5.8	5.7	5.6	7.0	7.8	8.7	9.9	9.5	10.4	11.3
DOE Equipment Deflated*	14.1	12.9	10.8	14.7	9.1	11.0	10.1	9.3	10.8	10.9	10.7	11.0	9.8	10.4	10.9
NSF Operating	6.5	9.2	11.2	12.5	13.2	16.3	16.9	20.3	20.4	20.9	21.3	23.2	23.5	25.6	30.8
NSF Theory	1.0	0.6	1.0	1.0	0.9	1.0	1.0	1.2	1.5	1.6	1.8	1.8	2.3	2.2	2.5
NSF Total Operating	7.5	9.8	12.2	13.5	14.1	17.3	17.9	21.5	21.9	22.5	23.1	25.0	25.8	27.8	33.3
NSF Total Operating Deflated*	19.6	24.3	29.2	31.1	29.8	32.9	31.8	35.5	33.8	31.5	28.4	27.7	26.7	27.8	32.3
Grand Total[a]	60.9	61.5	61.2	67.5	70.2	82.2	89.1	102.9	108.9	117.3	126.1	137.6	147.7	154.3	175.6
Grand Total Deflated	159.0	152.6	146.6	155.5	148.4	156.4	158.3	170.0	168.2	164.0	155.3	152.6	152.8	154.3	170.1
Construction															
DOE	19.9	15.4	14.4	4.7	0.6	1.9	7.5	10.7	4.6	10.4	14.3	11.3	12.0	7.7	13.7
NSF	1.2	0.2	0.1	0.0	0.2	0.2	0.8	0.7	1.2	1.3	1.6	1.9	2.1	8.9	8.7
Total	21.1	15.6	14.5	4.7	0.8	2.1	8.3	11.4	5.8	11.7	15.9	13.2	14.1	16.6	22.4
Total Deflated	55.1	38.7	34.7	10.8	1.7	4.0	14.7	18.8	9.0	16.4	19.6	14.6	14.6	16.6	21.7

NOTE: Graph is constructed from rows with asterisks.
[a] Grand total includes operating, theory, and equipment.

TABLE S3.6 Federal Obligations (Excluding Construction) for Subfields of Physics, Fiscal Years 1980-1984 (in Millions of Dollars)

	Fiscal Year				
	1980	1981	1982	1983	1984
Elementary-Particle Physics					
DOE Operations	207.2	235.0	255.6	308.6	324.6
DOE Theory	10.7	12.4	13.6	17.2	17.9
DOE Equipment	36.0	37.5	40.7	47.5	51.5
DOE Subtotal	253.9	284.9	309.9	373.3	394.0
NSF Operations + Equipment	22.6	25.2	26.2	28.7	35.9
NSF Theory	3.7	4.5	4.7	5.2	6.0
NSF Subtotal	26.3	29.7	30.9	33.9	41.9
Total Current	280.2	314.6	340.8	407.2	435.9
Total Deflated (CPI-W)	345.1	348.9	352.5	407.2	422.2
Nuclear Physics					
DOE Operations	88.2	95.7	104.7	107.9	122.0
DOE Theory	6.1	7.0	7.7	8.2	9.0
DOE Equipment	8.7	9.9	9.5	10.4	11.3
DOE Subtotal	103.0	112.6	121.9	126.5	142.3
NSF Operations + Equipment	21.3	23.2	23.5	25.6	30.8
NSF Theory	1.8	1.8	2.3	2.2	2.5
NSF Subtotal	23.1	25.0	25.8	27.8	33.3
NASA Int. Energy	0.1	0.2	0.1	0.1	0.1
Total Current	126.2	137.8	147.8	154.4	175.7
Total Deflated (CPI-W)	155.4	152.8	152.9	154.4	170.2
Atomic, Molecular, and Optical Physics					
DOE	5.3	5.7	6.0	6.4	6.9
NSF	5.6	7.1	8.0	8.7	10.9
NASA	0.3	0.1	0.4	0.5	0.5
ONR	6.0	7.4	7.7	7.3	9.3
AFOSR	4.4	5.0	6.1	6.4	7.0
ARO	8.3	9.5	10.1	10.1	10.3
DOD Subtotal	18.7	21.9	23.9	23.8	26.6
Total Current	29.9	34.8	38.3	39.4	44.9
Total Deflated (CPI-W)	36.8	38.6	39.6	39.4	43.5
Condensed-Matter Physics					
DOE	34.9	38.9	40.6	46.0	50.3
NSF	30.7	34.6	36.2	36.4	43.7
NASA	0.3	0.1	0.4	0.5	0.5
ONR	11.4	14.5	16.0	17.2	19.9
AFOSR	1.5	1.8	1.9	2.0	2.2
ARO	5.8	7.9	8.4	8.9	9.1
DOD Subtotal	18.7	24.2	26.3	28.1	31.2
Total Current	84.6	97.8	103.5	111.0	125.7
Total Deflated (CPI-W)	104.2	108.5	107.1	111.0	121.8

TABLE S3.6 Continued

	Fiscal Year				
	1980	1981	1982	1983	1984
Plasma and Fluid Physics					
DOE	287.7	325.1	417.3	491.2	529.4
NSF	1.7	1.9	1.6	1.5	1.6
NASA	12.7	14.4	15.5	17.3	17.3
ONR	1.8	1.7	2.3	3.4	4.2
AFOSR	0.9	1.0	1.1	1.2	1.3
DOD Subtotal	2.7	2.7	3.4	4.6	5.5
Total Current	304.8	344.1	437.8	514.6	553.8
Total Deflated (CPI-W)	375.4	381.6	452.8	514.6	536.3
Gravitation, Cosmology, and Cosmic-Ray Physics					
NSF	3.2	3.8	4.5	4.8	6.0
NASA	1.4	1.3	1.9	2.3	2.3
Total Current	4.6	5.1	6.4	7.1	8.3
Total Deflated (CPI-W)	5.7	5.7	6.6	7.1	8.0

ments, for example; the increases over the past years are a continuation of real increases over the past decade, which have approximately tripled the funding support in this subfield since FY 1974. Although Figure S3.7 shows substantial real growth in the subfield of plasma and fluid physics, there is concern (discussed more specifically in the Plasma and Fluid Physics volume of this Survey) that because of the mission-oriented goals of a large part of this subfield, most of the growth is directed toward the applied and development aspects of the work, with the result that sound basic research programs are not keeping pace with the rest of the subfield. Although the approach may be effective in achieving short-term goals, it is clearly not a wise long-term policy.

The weakening of the interface between the Department of Defense (DOD) and basic physics research was triggered by the Mansfield Amendment in 1969. The reduction of DOD sponsorship for basic research had an effect on most of the subfields in physics, but it seems to have been particularly severe in atomic, molecular, and optical physics (~20 percent funding loss) because the initial DOD support had been such a large fraction (~33 percent) of atomic, molecular, and optical physics funding and because other agencies did not pick up many of those grants and contracts. Although the Mansfield Amendment was no longer in effect by 1971, the connections between basic scientific research and national defense have still not been re-established. This problem has also been compounded by the shift of the defense research agencies away from long-range research or even development, so that not only has the quantity of effort been reduced, but at the same time a smaller fraction of the reduced effort is directed toward long-range research. This is an important concern both to the basic science community and to the national defense community; and it is important for both communities that their interconnection be re-established as expeditiously as possible, not only to provide much-needed support for basic research to expand man's horizons, but, just as

ORGANIZATION AND SUPPORT OF PHYSICS   131

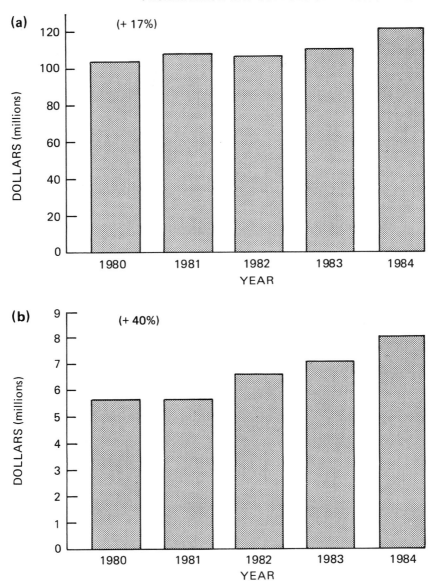

FIGURE S3.7 Federal funding support (excluding construction) for subfields of physics, FY 1980-FY 1984. This is the only period for which complete and consistent data could be extracted for all six subfields. Funding is expressed in FY 1983 dollars using CPI-W inflation factor. Percentages in parentheses are cumulative change in funding from FY 1980 to FY 1984. (a) Condensed matter; (b) cosmology, gravitation, and

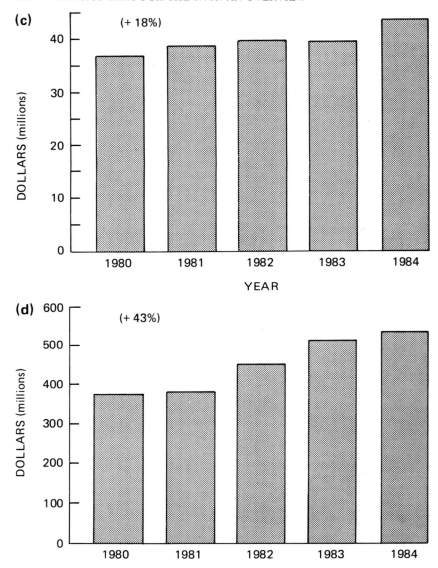

cosmic rays; (c) atomic, molecular, and optical; (d) plasmas and fluids; (e) elementary particles; (f) nuclear.

Note: Because of different definitions (e.g., what is "basic" versus what is "applied") the data in Figure S3.7, although internally consistent, do not add up to give the data in Figures S3.1 and/or Figure S3.3. [The data in Table S3.6 and Figure S3.7 were supplied

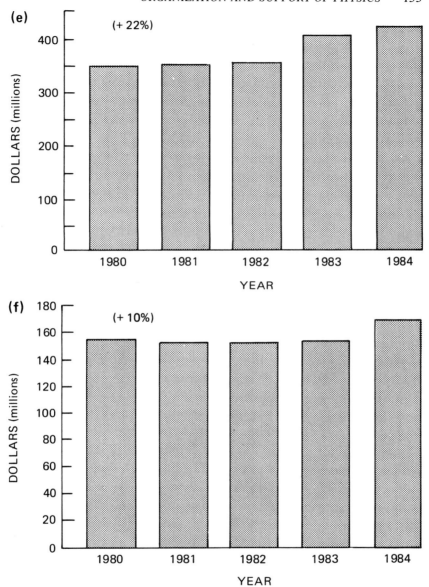

directly by the relevant program officers or administrators who are closer to the actual detailed disbursement of the funds than are the sources of Tables S3.1 and S3.2 (Figures S3.1 and S3.3, respectively).]

TABLE S3.7 Construction Projects

	Fiscal Year	Total Construction Cost (Millions of Dollars)
Plasma/Fusion Physics		
Inertial Fusion		
Shiva (LLNL)	74-76	25.0
Antares (LANL)	75-82	62.3
NOVA (LLNL)	78-82	176.0
Particle Beam Fusion Accelerator (SNL):		
PBFA I	75-78	14.2
PBFA II	81-85	45.7
Target Fabrication Facilities:		
(LANL)	80-82	15.3
(LLNL)	80-82	7.6
Magnetic Fusion		
Princeton Plasma Physics Laboratory (PPPL):		
Tokamak Fusion Test Reactor	76-84	313.6
PDX Neutral Beams	79-80	15.1
Mirror Fusion Test Facility (LLNL)	78-86	246.2
Doublet III Neutral Beams (General Atomic)	79-80	20.9
Elmo Bumpy Torus[a] (ORNL)	80-82	17.8
Elementary Particle Physics		
Fermi National Accelerator Laboratory (FNAL):		
200 GeV Accelerator	68-74	243.5
Energy Saver/Doubler	79-82	50.8
Tevatron I	81-86	82.5
Tevatron II	82-85	49.0
Stanford Linear Accelerator (SLAC):		
Positron-Electron Project (PEP)	76-83	80.3
SLAC Linear Collider (SLC)	84-86	112.0
Cornell Electron-Positron Storage Ring (Cornell)	77-79	19.6
Isabelle/CBA[a] (BNL)	78-83	124.0
Nuclear Physics		
SuperHILAC/Bevalac Upgrades (LBL)	70-80	12.4
Holifield Heavy-Ion Research Facility (ORNL):	75-79	17.2
National Superconducting Cyclotron Lab. (MSU):		
Phase I	75-79	2.9
Phase II	80-85	33.0
SUNY (Stony Brook)	77-82	4.1
Bates Linear Electron Accelerator (MIT):		
Experimental Facilities	77-79	5.0
Beam Recirculator	80-83	1.9
Argonne Tandem-Linac (ATLAS) (ANL)	82-83	7.7
Florida State University	83-85	2.8
Indiana University Cyclotron Facility	83-86	6.0
University of Washington	84-87	8.0
Yale University	84-87	11.0

TABLE S3.7 Continued

	Fiscal Year	Total Construction Cost (Millions of Dollars)
Multidisciplinary Facilities		
National Synchrotron Light Source (BNL)	78-80	24.0
Intense Pulsed Neutron Source (ANL)	79-80	8.8
Stanford Synchrotron Radiation Lab (SSRL)	73-80	7.9
Aladdin Synchrotron Radiation Center (SRC)	77-80	3.5

[a] Canceled.

important, to provide expert advice and communication between these sectors and to provide new ideas and concepts for future technological developments.

Construction fund budgets are subject to large fluctuations in response to other pressure in the federal budget process. For example, the budget in 1974 was ~$15 million compared with ~$190 million in 1979. For this reason, it is not useful to plot construction funding on a year-by-year basis. Instead, in Table S3.7 we have listed the major facility construction projects over the past 15 years together with the period of their construction and their total cost in current (i.e., actual) dollars.

Several large new projects for construction during the next 10-12 years are in the planning and proposal stages. These are listed earlier in Table 3.2; their characteristics are discussed in the appropriate panel reports of this Survey. The BCX facility is proposed as the next major step in magnetic fusion research beyond the TFTR tokamak. The SSC, CEBAF, and RNC accelerator facilities are proposed as the next steps in upgrading the energies available for elementary-particle physics and nuclear physics, respectively.

ORGANIZATION AND DECISION MAKING

As each of the subfields of physics continues to evolve toward more centralized big science, the specific decisions made by funding organizations have a larger and more direct effect on the subfield. In a broad research program, each group inevitably tends to regard its own efforts as undersupported relative to their importance, and thus there is constant pressure to reallocate resources. The simplest administrative response is to maintain the status quo, and it requires intelligent, objective oversight to recommend constructive changes. There is, therefore, a need for the individual subfields to take on more active and responsible roles in developing a consensus within the subfield itself about priorities and directions of development in order to provide the funding organizations with the informed critical advice that they need to make allocation decisions.

At the highest level this involves advisory groups such as the White House Science Council, the National Science Board (NSF), and the Energy Research Advisory Board (DOE). Each of the funding agencies also frequently makes use of ad hoc advisory committees related to specific decisions affecting specific subfields. For example, beginning in the 1950s with the need for making decisions about new accelerator facilities at MURA, Argonne, and Stanford, the government found it necessary to seek advice from so many ad hoc advisory panels in the area of elementary-particle physics that, in 1967, the AEC formed the High Energy Physics Advisory Panel (HEPAP) as a standing committee to provide advice and expertise regarding the issues to be confronted in making decisions in this subfield. Ad hoc panels and committees are still frequently formed when necessary, but these are now generally advisory to HEPAP, which considers reports and then funnels its own conclusions to the agencies. It is also worth noting that the European particle-physics community has analogous institutions. CERN is governed by a council comprising scientific and political representatives from the CERN member nations. A Scientific Policy Committee is advisory to the CERN administration. In addition, there is a standing European Committee on Future Accelerators (ECFA), which considers long-range planning issues for Europe.

In 1977, following one of the recommendations of the NRC Ad Hoc Panel on the Future of Nuclear Science in its report *Future of Nuclear Science* (National Academy of Sciences, Washington, D.C., 1977, p. 79):

In a frontier field such as nuclear science, priorities and directions change rapidly with time, new discoveries spawn new programs and may make old ones obsolete, and funding constraints are not immutable. All these factors make it essential that a Nuclear Science Advisory Panel be established to advise the funding agencies on a continuing rather than an *ad hoc* basis. . . . We believe that nuclear science has suffered by not having had an active body to advise, on a continuing basis, on the needs and status of the field and on the balance between various programs.

ERDA and the NSF established the Nuclear Science Advisory Committee (NSAC) to play the same role for the subfield of nuclear physics that HEPAP plays for elementary-particle physics. More recently a similar Magnetic Fusion Advisory Committee (MFAC) has also been formed.

Advisory committees such as these can provide funding agencies with evaluations and recommendations concerning both the long-range objectives and priorities of their subfield and its specific needs for funding, manpower, instrumentation, and facilities. It seems clear, from our discussions within the Steering Committee for the Physics Survey, that the other areas and subfields of physics would benefit by having an active body to advise, on a continuing basis, on the needs and status of the subfield and on the balance between various programs. It is also important (on a broader scale) for the various Divisions of the American Physical Society (APS) to monitor the needs of their subfields as well as any changes in the budgets of the various funding agencies and to develop their own expertise in analyzing and understanding such information. It is a healthy sign that in some subfields the respective APS Divisions have started to take more active roles in policy issues.

In comparing research in Europe and in the United States, the 1982 NRC report *Outlook for Science and Technology: The Next Five Years* (Chapter 13) notes that one organizational difference between Europe and the United States is that the European countries generally have many more centralized funding and decision-making processes. This can have the advantage of providing greater funding stability (something that many research groups in this country would like to see), but at the same time it has the disadvantage of reducing the flexibility and diversity of the system. By removing decisions from the local operating level, the more centralized European mode has the disadvantage of reducing its ability to respond readily to new, unexpected opportunities. In the trade-off between greater budgetary stability and flexibility, the U.S. system shows greater diversity and competitiveness, which gives us a greater ability to respond to new research opportunities and provides a wider range of opportunities for young scientists.

Abbreviations and Acronyms

AAUP	American Association of University Professors
AEC	Atomic Energy Commission
AFOSR	Air Force Office of Scientific Research
AGS	Alternating Gradient Synchrotron
AIP	American Institute of Physics
AMO	atomic, molecular, and optical (physics)
ANL	Argonne National Laboratory
APS	American Physical Society
ARO	Army Research Office
ATLAS	Argonne Tandem Linear Accelerator System
BCX	Burning Core Experiment (plasma physics)
BLS	Bureau of Labor Statistics
BNL	Brookhaven National Laboratory
CARS	coherent anti-Stokes Raman scattering
CBA	Colliding-Beam Accelerator (formerly Isabelle)
CEA	Cambridge Electron Accelerator
CEBAF	Continuous Electron Beam Accelerator Facility
CERN	Conseil Européen de la Recherche Nucléaire
CESR	Cornell Electron-Positron Storage Ring
CM	condensed matter
CMP	condensed-matter physics
CPI-W	Consumer Price Index-Urban and Clerical Workers (BLS)
DESY	Deutsches Electronen Synchrontron

DNA	deoxyribonucleic acid
DOD	Department of Defense
DOE	Department of Energy
ECFA	European Committee on Future Accelerators
EP	elementary particle
EPP	elementary-particle physics
ERDA	Energy Research and Development Administration
EXAFS	extended x-ray absorption fine structure
eV	electron volt
FNAL	Fermi National Accelerator Laboratory
GeV	billion electron volts, 1×10^9 electron volts, giga electron volt
HEPAP	High Energy Physics Advisory Panel
JET	Joint European Torus (plasma project)
keV	thousand electron volts, 1×10^3 electron volts, kilo electron volt
LANL	Los Alamos National Laboratory
LBL	Lawrence Berkeley Laboratory
LLNL	Lawrence Livermore National Laboratory
MeV	million electron volts, 1×10^6 electron volts, mega electron volt
MFAC	Magnetic Fusion Advisory Committee
MFECC	Magnetic Fusion Energy Computer Center
MIT	Massachusetts Institute of Technology
MOCVD	metal-organic chemical-vapor deposition
MRI	Magnetic Resonance Imaging
MSU	Michigan State University
MURA	Midwestern Universities Research Association
NAS	National Academy of Sciences
NASA	National Aeronautics and Space Administration
NBS	National Bureau of Standards
NCAR	National Center for Atmospheric Research (NSF)
NIE	National Institute for Education
nm	nanometer, 1×10^{-9} meter
NRC	National Research Council
NSAC	Nuclear Science Advisory Committee
NSF	National Science Foundation
ONR	Office of Naval Research
ORNL	Oak Ridge National Laboratory
PEP	Positron-Electron Project
PPA	Princeton-Pennsylvania Accelerator
PPPL	Princeton Plasma Physics Laboratory

RNC	Relativistic Nuclear Collider
SLAC	Stanford Linear Accelerator Center
SLC	Stanford Linear Collider
SNL	Sandia National Laboratories
SPEAR	SLAC Positron-Electron Asymmetric Ring
SRC	Sychrotron Radiation Center
SSC	Superconducting Super Collider
SSRL	Stanford Synchrotron Radiation Laboratory
TeV	trillion electron volts, 1×10^{12} electron volts, tera electron volt
TFTR	Tokamak Fusion Test Reactor
TIAA-CREF	Teachers Income Annuity Association-College Retirement Equities Fund
ZGS	Zero Gradient Synchrotron

Glossary of Physical Terms

Absolute zero. The temperature of $-273.16°C$, or $-459.69°F$, or zero kelvin (0 K), thought to be the temperature at which molecular motion is at a minimum and a body has no heat energy.

Alpha particle. A positively *charged particle* consisting of two *protons* and two *neutrons*, identical with the *nucleus* of the helium *atom*; emitted by several radioactive substances.

Angstrom. A unit of length equal to one ten-billionth of a meter (10^{-10} m).

Anisotropy. The characteristic of a substance for which a physical property varies in value with the direction in or along which the measurement is made.

Antimatter. Material consisting of *atoms* that are composed of *positrons, antiprotons,* and *antineutrons*.

Antineutron. The *antiparticle* to the *neutron*; a strongly interacting *baryon* that has no charge, mass of 939.6 MeV, *spin* 1/2, and mean life of almost 10^3 seconds.

Antinucleon. An *antineutron* or *antiproton*, that is, a particle having the same mass as its *nucleon* counterpart but opposite charge or opposite *magnetic moment*.

Antiparticle. A counterpart to a particle, having mass, lifetime, and *spin* identical to the particle but with charge and *magnetic moment* reversed in sign.

Antiproton. The *antiparticle* of the *proton*, a strongly interacting *baryon* that is stable, carries unit negative charge, has the same mass as the *proton* (983.3 MeV), and has *spin* 1/2.

Astrophysics. The study of such physical properties of celestial bodies as *luminosity*, size, mass, density, temperature, and chemical composition; the study of the origin and evolution of these bodies.

Asymptotic freedom. In some *particle-physics* theories, the binding force between two *quarks* decreases as their relative momentum increases; equivalently, as two *quarks* approach, the force between them disappears.

Atom. The individual structure that constitutes the basic unit of any chemical element.

Atomic number. The number of *protons* in an atomic *nucleus*.

Atomic physics. The science concerned with the structure of the *atom*. The characteristics of the *elementary particles* of which the *atom* is composed and the processes involved in the interactions of radiant energy with *matter*.

Atomic spectrum. The spectrum of radiations due to transitions between energy levels in an *atom*, either absorption or emission.

Aurora. The most intense of the several *lights* emitted by the Earth's upper atmosphere, seen most often along the outer realms of the Arctic and Antarctic, where it is called the aurora borealis and aurora australis, respectively; excited by *charged particles* from space.

Baryon. A particle that can be transformed into a *nucleon* and some number of *mesons* and lighter particles; any of a group of *hadrons* (as *nucleons*) that undergo *strong interactions* and are held to be a combination of three *quarks*.

Beta decay. Radioactive transformation of a *nuclide* in which the *atomic number* increases or decreases by unity with no change in *mass number*; the *nucleus* emits a *beta particle*.

Beta particle. An *electron* or *positron* emitted from a *nucleus* during *beta decay*.

Big bang. A theory in astronomy, according to which the universe originated billions of years ago from the explosion of a single mass of material, so that the pieces are still flying apart.

Biophysics. The hybrid science involving the application of physical principles and methods to study and explain the structures of living organisms and the mechanics of life processes.

Black hole. A star with radius just outside the *Schwarzschild radius*; it is invisible but can capture *matter* and *light* from the outside.

Boson. A particle (as a *photon*, *meson*, or *alpha particle*) whose *spin* is zero or an integral number.

"Breathing" mode. The vibrational state in which a *nucleus* undergoes spherically symmetric radial expansion and contraction.

Charged particle. A particle whose charge is not zero; the charge of a particle is added to its designation as a superscript with particles of $+1$ and -1 (in terms of the charge of the *proton*) denoted by $+$ and $-$, respectively.

Coherent anti-Stokes Raman scattering. Nonlinear spectroscopy where *laser light* scattered from a sample undergoes an increase in frequency (anti-Stokes behavior), derived from vibrational states of the *atoms* or *molecules* in the sample.

Condensed matter (physics). The physics of the solid and liquid states.

Cosmic rays. Electrons, muons, the nuclei of *atoms*, and *photons* that impinge upon the Earth from all directions of space with nearly the speed of *light*.

Cosmology. The study of the overall structure of the physical universe.

Crystallography. The branch of science that deals with the geometric description of crystals and their internal arrangement.

Cyclotron. An accelerator in which *charged particles* are successively accelerated by a constant-frequency alternating electric field that is synchronized with movement of the particles on spiral paths in a constant magnetic field normal to their path.

Deuteron. The *nucleus* of the deuterium *atom* consisting of one *proton* and one *neutron*.

Diabatic. A change in which the environment of a system alters too rapidly for the system to read just continuously.

Diamagnetic. Having a *magnetic permeability* less than that of a vacuum; slightly repelled by a magnet.

Dirac electron theory. Theory that accounts for *spin* angular momentum of the *electron* and gives its *magnetic moment* and its behavior in an electromagnetic field.

Doppler effect. The change in the observed frequency of a *wave* due to relative motion of source and observer.

Electromagnetic theory. Theory according to which *light* is an electromagnetic *wave* whose electric and magnetic fields obey Maxwell's equations.

Electromagnetism. Magnetism produced by an electric current rather than by a permanent magnet.

Electron. An *elementary particle* consisting of a charge of negative electricity equal to about 1.602×10^{-19} coulomb and having a mass when at rest of about 9.109534×10^{-28} gram (or about 1/1836 that of a *proton*).

Electron accelerator. A device that accelerates *electrons* to high energies.

Electron charge. The charge carried by an *electron*, equal to about -1.602×10^{-19} coulomb, or -4.803×10^{-10} statcoulomb.

Elementary particle. A particle that, in the present state of knowledge, cannot be described as compound and is thus one of the fundamental constituents of all *matter*.

Entropy. The degradation of the *matter* and energy in the universe to an ultimate state of inert uniformity.

Epitaxy. Growth of one crystal on the surface of another crystal, in which the growth of the deposited crystal is oriented by the lattice structure of the *substrate*.

Equivalence Principle. In *general relativity*, the principle that the observable local effects of a gravitational field are indistinguishable from those arising from acceleration of the frame of reference.

Far-ultraviolet radiation. Ultraviolet radiation in the wavelength of 200 to 300 nanometers; germicidal effects are greatest in this range.

Femto-. Prefix meaning one quadrillionth (10^{-15}) part of.

Fermion. A particle (as an *electron*, *proton*, or *neutron*) whose *spin quantum* number is an odd multiple of 1/2.

Ferromagnet. A substance with an abnormally high *magnetic permeability*, a definite saturation point, and appreciable residual magnetism and hysteresis.

Fission. The division of an atomic *nucleus* into parts of comparable mass; usually restricted to heavier nuclei such as isotopes of uranium, plutonium, and thorium.

Fluorescence. Emission of, or the property of emitting, electromagnetic radiation (usually as visible *light*) resulting from and occurring only during absorption of radiation from some other source.

Free-electron laser. A *laser* in which beams of unbound *electrons* interact with a strong magnetic field to produce tunable *laser light*.

Free radical. An *atom* or a diatomic or polyatomic *molecule* that possesses at least one unpaired *electron*.

Fusion. Combination of two light nuclei to form a heavier *nucleus* (and perhaps other reaction products) with release of some binding energy.

Galaxy. A large-scale aggregate of stars, gas, and dust. The aggregate is a separate system of stars covering a mass range from 10^7 to 10^{12} *solar masses* and ranging in diameter from 1500 to 300,000 *light-years*.

Gamma ray. A high-energy *photon*, especially as emitted by a *nucleus* in a transition between two energy levels.

General relativity. The theory of Einstein that generalizes special relativity to noninertial frames of reference and incorporates *gravitation* and in which events take place in a curved space.

Geophysics. The physics of the Earth and its environment, i.e., earth, air, and (by extension) space.

Glass. A hard, amorphous, inorganic, usually transparent, brittle substance made by fusing silicates, sometimes borates and phosphates, with certain basic oxides and then rapidly cooling to prevent crystallization (NB: not the sense in which the term is used in the term *spin glass*).

Gluon. A hypothetical, neutral, massless particle believed to bind together *quarks* to form *hadrons*.

Gluon string. A *particle-physics* theoretical model to account for the binding force between *quarks* that increases monotonically as they are separated.

Gravitation. The mutual attraction among all masses in the universe.

Gravitational force. The force on a particle due to its gravitational attraction to other particles.

Gravitational radiation. A propagating gravitational field predicted by *general relativity*, which is produced by some change in the distribution of *matter*; it travels at the speed of *light*, exerting forces on masses in its path.

Gravitational redshift. A displacement of spectral lines toward the red when the gravitational potential at the observer of the *light* is greater than at its source.

Gyrotron. A device for producing microwave energy. Also called an electron cyclotron *maser*.

Hadron. Any of the particles that take part in the *strong interaction*.

Hall conductivity. The reciprocal of the electrical resistivity associated with the *Hall current*.

Hall current. When an electric current in a conductor is placed in a magnetic field that is perpendicular to the current a transverse electric field is created, which in turn can support a transverse current.

Heavy-ion linear accelerator. A linear accelerator that produces a beam of heavy particles of high intensity and sharp energy; used to produce transuranic elements and short-lived isotopes and to study nuclear reactions, nuclear *spectroscopy*, and the absorption of heavy ions in *matter*.

Helicity. The component of the *spin* of a particle along its momentum.

Hydrogen maser. A *maser* in which hydrogen gas is the basis for providing an output signal with a high degree of stability and spectral purity.

Hypercharge. A *quantum* number conserved by *strong interactions*, equal to twice the average of the changes of the numbers of an *isospin multiplet*.

Hypernucleus. A *nucleus* containing one or more *hyperons* in addition to the *nucleons*.

Hyperon. A *hadron* that has *baryon* number $B = +1$, i.e., that can be transformed into a *nucleon* and some number of *mesons* or lighter particles and that has a nonzero strangeness number.

Ionicity. The state of being characterized by, relating to, or existing as ions.

Isospin multiplet. A collection of *hadrons* that have approximately the same mass and the same *quantum* numbers except for charge.

Isotropy. The quality of a property that does not depend on the direction along which it is measured or of a medium or entity whose properties do not depend on the direction along which they are measured.

Joule. Unit of energy, work, or quantity of heat equal to one *newton*-meter.

J/ψ particle. An unstable, neutral *meson* that has a mass about 3 times the mass of the *photon*.

K meson. See kaon.

Kaon. Collective name for four pseudoscalar *mesons* having masses of about 495 MeV and decaying via *weak interactions*; an unstable *meson* produced in high-energy particle collisions with its electrically charged forms being 966.3 times more massive than the *electron*; also known as *K meson*.

Lamb shift. A small shift in the energy levels of a hydrogen atom, and of hydrogenlike ions, from those predicted by the *Dirac electron theory*, in accord with principles of *quantum electrodynamics*.

Lambda hyperon (Λ). A quasi-stable *baryon*, forming an isotopic singlet, having zero charge and *hypercharge*, a *spin* of 1/2, positive *parity*, and mass of 1115.5 MeV.

Laser. A device that uses the *maser* principle of amplification of electromagnetic *waves* by stimulated emission of radiation and operates in the optical or infrared region. Derived from *l*ight *a*mplification by *s*timulated *e*mission of *r*adiation.

Laser interferometer. An interferometer that uses a *laser* as a *light* source; because of the monochromaticity and high intrinsic brilliance of *laser light*, it can operate with path differences in the interfering beams of hundreds of *meters*, in contrast to a maximum of about 20 centimeters for classical interferometers.

Laser light. Coherent, nearly single-frequency, highly directional electromagnetic radiation emitted in the range from infrared to ultraviolet and x-ray wavelengths.

Laser optics. Optical systems utilizing the properties of *laser light.*

Laser spectroscopy. A branch of *spectroscopy* in which a *laser* is used as an intense, monochromatic *light* source; in particular, it includes saturation *spectroscopy,* as well as the application of *laser* sources to *Raman spectroscopy* and other techniques.

Lepton. Any of a family of particles (as *electrons, muons,* and *neutrinos*) that have *spin* quantum number 1/2 and that experience no *strong interactions*; a *fermion* having a mass smaller than the *proton* mass.

Light. Electromagnetic radiation with wavelengths capable of causing the sensation of vision, ranging approximately from 4000 (extreme violet) to 7700 *angstroms* (extreme red); more generally, electromagnetic radiation of any wavelength.

Light-year. A unit of measurement of astronomical distance; it is the distance *light* travels in one *sidereal year* and is equivalent to 9.461×10^{12} kilometers or 5.879×10^{12} miles.

Luminosity. In optics, a measure of the brightness of a light source; in colliding-beam accelerators, a measure of the rate of collisions of the particles in the colliding beams.

Macromolecule. A large *molecule* in which there is a large number of one or several relatively simple structural units, each consisting of several *atoms* bonded together.

Magnetic hysteresis. Lagging of changes in the magnetization of a substance behind changes in the magnetic field as the magnetic field is varied.

Magnetic moment. A vector associated with a magnet, current loop, particle, or such, whose cross product with the magnetic induction (or alternatively, the magnetic-field strength) of a magnetic field is equal to the torque exerted on the system by the field.

Magnetic monopole. A hypothetical particle carrying magnetic charge; it would be a source for magnetic field in the same way that a *charged particle* is a source for electric field.

Magnetic permeability. A factor, characteristic of a material, that is proportional to the magnetic induction produced in a material divided by the magnetic-field strength; it is a tensor when these qualities are not parallel.

Magnetic resonance. A phenomenon exhibited by the magnetic *spin* systems of certain *atoms* whereby the *spin* systems absorb energy at specific (resonant) frequencies when subjected to magnetic fields alternating at frequencies that are in synchronism with natural frequencies of the system.

Maser. A device for coherent amplification or generation of electromagnetic *waves* in which an ensemble of *atoms* or *molecules*, raised to an unstable energy state, is stimulated by an electromagnetic *wave* to radiate excess energy at the same frequency and *phase* as the stimulating *wave*.

Mass number. The sum of the numbers of *protons* and *neutrons* in the *nucleus* of an *atom* or *nuclide*.

Matter. The substance composing bodies perceptible to the senses; includes any entity possessing mass when at rest.

Meson. Any of a group of *hadrons* (as the *pion* and *kaon*) that are strongly interacting and have zero or an integer number of *quantum* units of *spin*.

Metastability. The property of having only a slight margin of stability.

Metrology. The science of weights and measures or of measurement.

Mho. A unit of conductance, admittance, and susceptance equal to the conductance between two points of a conductor such that a potential difference of 1 volt between these points produces a current of 1 ampere; the conductance of a conductor in mhos is the reciprocal of its resistance in ohms.

Microelectronics. The technology of constructing circuits and devices in extremely small packages by various techniques.

Microemulsion. A homogenous, single-*phase*, thermodynamically stable mixture of oil, water, and surfactant.

Million electron volts (MeV). A unit of energy commonly used in nuclear and *particle physics*, equal to the energy acquired by an *electron* in falling through a potential of 10^6 volts.

Molecular biology. That part of biology which attempts to interpret biological events in terms of the physicochemical properties of *molecules* in a cell.

Molecular ion. A *molecule* possessing nonzero net electric charge.

Molecular physics. The study of the behavior and structure of *molecules*, including the *quantum-mechanical* explanation of several kinds of chemical binding between *atoms* in a *molecule*; directed valence; the polarizability of *molecules*; the quantization of vibrational, rotational, and electronic motions of *molecules*; and the phenomena arising from intermolecular forces.

Molecule. A group of atoms held together by chemical forms; a molecule is the smallest unit of *matter* that can exist by itself and retain all its chemical properties.

Monte Carlo method. A technique that obtains a probabilistic approximation to the solution of a problem by using statistical sampling techniques.

Multiphoton spectroscopy. Nonlinear spectroscopy usually involving intense *laser light* enabling more than one *photon* to interact with an *atom* within a time frame shorter than the decay time of the atomic state.

Muon. Collective name for two semistable *elementary particles* with positive and negative charge, which are *leptons* and have a *spin* of 1/2 and a mass of approximately 105.7 MeV.

Muonium. An *atom* consisting of an *electron* bound to a positively charged *muon* by their mutual Coulomb attraction, just as an *electron* is bound to a *proton* in the hydrogen *atom*.

Nano-. Prefix meaning one billionth (10^{-9}) part of.

Neutrino. A neutral *lepton* having zero rest mass and *spin* 1/2.

Neutron. An uncharged *hadron* that has a mass nearly equal to that of the *proton* and is present in all known atomic nuclei except the hydrogen *nucleus*.

Newton. The unit of force equal to one kilogram-meter per second squared (1 kg · m/s^2).

Nonlinear optics. The study of the interaction of radiation with *matter* in which certain variables describing the response of the *matter* (such as electric polarization of power absorption) are not proportional to variables describing the radiation (such as electric-field strength or energy flux).

Nonlinear spectroscopy. *Spectroscopy* usually using *laser light* and the *nonlinear optical* behavior of certain properties of matter; e.g., *coherent anti-Stokes Raman scattering*.

Nova. A star that suddenly becomes explosively bright (the term is a misnomer because it does not denote a new star but the brightening of an existing faint star).

Nuclear physics. The study of the characteristics, behavior, and internal structures of the atomic *nucleus*.

Nucleon. A collective name for a *proton* or a *neutron*; these particles are the main constituents of atomic nuclei, have approximately the same mass, have a *spin* of 1/2, and can transform into each other through the process of *beta decay*.

Nucleosynthesis. The production of a chemical element from hydrogen nuclei (as in stellar evolution).

Nucleus. The central, positively charged, dense portion of an *atom*.

Nuclide. A species of *atom* characterized by the number of *protons*, number of *neutrons*, and energy content in the *nucleus*; to be regarded as a distinct nuclide, the *atom* must be capable of existing for a measurable lifetime, generally greater than 10^{-12} second.

Parity. A physical property of a *wave* function that specifies its three behavior under an inversion, i.e., under simultaneous reflection of all three spatial coordinates through the origin; if the *wave* function is unchanged by inversion, its parity is 1 (or even); if the function is changed only in sign, its parity is −1 (or odd).

Particle physics. The branch of physics concerned with understanding the properties and behavior of *elementary particles*, especially through study of collisions or decays involving energies of hundreds of MeV or more.

Pauli Exclusion Principle. The principle that no two *fermions* of the same kind may simultaneously occupy the same *quantum* state.

Phase. A portion of a physical system (liquid, gas, *solid*) that is homogeneous throughout, has definable boundaries, and can be separated physically from other phases; the type of state of a system, such as *solid*, liquid, or gas.

Phase transition. A change of a substance from one *phase* (e.g., *solid*, liquid, or gas) to another.

Phonon. A *quantum* of an acoustic mode of thermal vibration in a crystal lattice.

Photoemission. The ejection of *electrons* from a *solid* (or less commonly, a liquid) by incident electromagnetic radiation.

Photoionization. The removal of one or more *electrons* from an *atom* or *molecule* by absorption of a *photon* of visible or ultraviolet *light*.

Photon. A massless particle, the *quantum* of the electromagnetic field, carrying energy, momentum, and angular momentum.

Photovoltaic. Of, relating to, or utilizing the generation of a voltage when radiant energy falls on the boundary between two dissimilar substances (as two different *semiconductors*).

Pion. A short-lived *meson* that is primarily responsible for the nuclear force and that exists as a positive or negative particle with mass 273.2 times the *electron* mass or as a neutral particle with mass 264.2 times the *electron* mass.

Planck's constant. A fundamental physical constant, the elementary *quantum* of action; the ratio of the energy of a *photon* to its frequency, it is equal to $6.62620 \pm 0.00005 \times 10^{-34}$ *joule*-second; symbolized by h.

Plasma. A collection of *charged particles* (as in the atmosphere of stars or in a metal) containing about equal numbers of positive ions and *electrons*, exhibiting some properties of a gas but differing from a gas in being a good conductor of electricity and in being affected by a magnetic field.

Plasma physics. The study of highly ionized gases.

Plate tectonics. Global *tectonics* based on a model of the Earth characterized by a small number (10-25) of semirigid plates that float on some viscous underlayer in the mantle; each plate moves more or less independently and grinds against the others, concentrating more deformation, volcanism, and seismic activity along the periphery.

Positive energy theorem. A recent theory that shows that in *general relativity* theory, any isolated system must have a positive value for its total energy.

Positron. An *elementary particle* having mass equal to that of the *electron* and having the same *spin* and statistics as the *electron* but a positive charge equal in magnitude to the *electron's* negative charge; the *antiparticle* of the *electron*.

Positronium. The bound state of an *electron* and a *positron*.

Proton. A *hadron* that is the positively charged constituent of ordinary *matter* and, together with the *neutron*, is a building stone of all atomic nuclei; its mass is approximately 938 MeV.

Proton accelerator. A particle accelerator that accelerates *protons* to high energies, as opposed to one that accelerates heavier ions or *electrons*.

Pulsar. A celestial radio source, emitting intense short bursts of radio emission; the periods of known pulsars range between 33 milliseconds and 3.75 seconds, and pulse durations range from 2 to about 150 milliseconds, with longer-period pulsars generally having a longer pulse duration.

Quantize. To restrict an observable quantity, such as energy or angular momentum, to a discrete set of values, to subdivide (as energy) into small but measureable increments.

Quantized Hall effect. The appearance of *quantum* levels in the *Hall conductivity* for a two-dimensional conductor in a magnetic field.

Quantum. For certain physical quantities, a unit such that the values of the quantity are restricted to integral multiples of this unit (e.g., the quantum of angular momentum is *Planck's constant* divided by 2); an entity resulting from quantization of a field or *wave*, having particlelike properties such as energy, mass, momentum, and angular momentum (e.g., the *photon* is the quantum of an electromagnetic field, and the *phonon* is the quantum of a lattice vibration).

Quantum chromodynamics. The *quantum* theory that describes the *strong interactions* that bind *quarks* together to form *hadrons*.

Quantum electrodynamics. The *quantum* theory of electromagnetic radiation, synthesizing the *wave* and corpuscular pictures, and of the interaction of radiation with electrically charged *matter*, in particular with *atoms* and their constituent *electrons*.

Quantum mechanics. The modern theory of *matter*, of electromagnetic radiation, and of the interaction between *matter* and radiation; it differs from classical physics, which it generalizes and supersedes, mainly in the realm of atomic and subatomic phenomena.

GLOSSARY OF PHYSICAL TERMS 153

Quark. Hypothetical *elementary particles* that have charges whose magnitudes are 1/3 or 2/3 of the *electron charge*; quarks are thought to come in several types (as up, down, strange, charmed, and bottom) and are held to be a constituent of *hadrons*.

Quark-gluon model. The *particle-physics* model of the constituents of *hadrons* and the force that binds them; see *quantum chromodynamics*.

Radiation pressure. The pressure exerted by electromagnetic radiation on objects on which it impinges.

Radio astronomy. The study of celestial objects by measurement and analysis of their emitted electromagnetic radiation in the wavelength range from roughly 1 millimeter to 30 meters.

Raman spectroscopy. Nonlinear *spectroscopy* named for Sir Chandrasekhara Venkata Raman (1888-1970), Indian physicist.

Relativistic magnetron. A device to produce microwaves that uses *electrons* moving at velocities near the speed of light in a magnetic field.

Renormalization group approach or theory. A mathematical technique to avoid infinities that occur in certain classes of physical theories.

Resistivity. The electrical resistance offered by a material to the flow of current times the cross-sectional area of current flow and per unit length of current path; the reciprocal of the conductivity.

Robotics. Technology dealing with the design, construction, and operation of robots in automation.

Schwarzschild radius. For a given body of *matter*, a distance equal to the mass of the body times the gravitational constant divided by the square of the speed of light.

Second-order Doppler effect. At velocities close to the speed of light, additional *Doppler effects* may be detected even in cases where the source and observer are moving only transversely.

Semiconductor. A *solid* crystalline material whose electrical conductivity is intermediate between that of a metal and an insulator, ranging from about 10^5 *mhos* to 10^{-7} *mho* per meter, and is usually strongly temperature dependent.

Semiconductor laser. A *laser* in which the wavelength of the coherent *light* beam is determined by a *semiconductor* compound.

Sidereal year. The time period relative to the stars of one revolution of the Earth about the Sun; it is about 365.2564 mean solar days.

Sigma-zero (Σ^0) particle. An unstable *elementary particle* of the *baryon* family, of neutral charge, with a mass about 1.4 times the mass of the *proton*.

Soft x ray. An x ray having a comparatively long wavelength and poor penetrating power.

Solar mass. The mass of the Sun, 2×10^{30} kg.

Solar physics. The scientific study of all physical phenomena connected with the Sun; it overlaps with *geophysics* in the consideration of solar-terrestrial relationships such as the connection between solar activity and *auroras*.

Solid. A substance that has a definite volume and shape and resists forces that tend to alter its volume or shape; a crystalline material, i.e., one in which the constituent *atoms* are arranged in a three-dimensional lattice, periodic in three independent directions.

Soliton(s). Solitary *waves* (as in a gaseous *plasma*) that retain their *phase* and speed after colliding with each other.

Space charge. An electric charge distributed throughout a three-dimensional region.

Spectroscopy. The branch of physics concerned with the production, measurement, and interpretation of electromagnetic spectra arising from either emission or absorption of radiant energy by various substances.

Spin. The intrinsic angular momentum of a particle or *nucleus*, which exists even when the particle is at rest, as distinguished from orbital angular momentum.

Spin glass. A state of matter in which the magnetic *spins* of randomly located atoms freeze in direction at low temperature; often generalized to other systems.

Strong force. See *strong interaction*.

Strong interaction. One of the fundamental interactions of *elementary particles*, primarily responsible for nuclear forces and other interactions among *hadrons*.

Substrate. The physical material on which a microcircuit is fabricated.

Superconducting magnet. An electromagnet whose coils are made of a superconductor with a high transition temperature and extremely high critical field; it is capable of generating magnetic fields of 100,000 oersteds and more with no steady power dissipation.

Superconductivity. A property of many metals, alloys, and chemical compounds at temperatures near *absolute zero* by virtue of which their electrical *resistivity* vanishes and they become strongly *diamagnetic*.

Superdense. Densities that are greater than that for an ordinary *nucleus*, such as may exist in the core of *novae*.

Supernova. A star that suddenly bursts into very great brilliance as a result of its blowing up; it is orders of magnitude brighter than a *nova*.

Supersymmetry. A *particle-physics* theory that attempts to unite two particle classes of *fermion*s and *boson*s into a unified theory.

Symmetry. The property of remaining invariant under certain changes (as of orientation in space, of the sign of the electric charge, of *parity*, or of the direction of time flow).

Synchrotron. A device for accelerating *electrons* or *protons* in closed orbits in which the frequency of the accelerating voltage is varied (or held constant in the case of *electrons*) and the strength of the magnetic field is varied so as to keep the orbit radius constant.

Tau particle [*or tau lepton* (τ)]. A short-lived *elementary particle* of the *lepton* family that exists in positive and negative charge states and has a mass about 3500 times heavier than an *electron*.

Tectonics. A branch of geology concerned with structure, especially with folding and faulting.

Tera-. Prefix meaning one trillion (10^{12}).

Tesla. Unit of magnetic flux intensity equal to one weber per square meter (1 Wb/m^2), or one volt second per square meter (1 V · s/m^2).

Three-degree radiation. The remnant radiation, at microwave frequencies, of the *big bang*.

Tokamak. A device for confining *plasma* within a *toroidal* chamber, which produces *plasma* temperatures, densities, and confinement times greater than those produced by any other such device.

Tomography. A diagnostic technique using x-ray photographs in which the shadows of structures before and behind the section under scrutiny do not show.

Toroidal. Of, relating to, or shaped like a torus; doughnut-shaped.

Upsilon particle (Υ). Any of a group of unstable electrically neutral *mesons* that have a mass about 10 times that of a *proton*.

Vacuum polarization. A process in which an electromagnetic field gives rise to virtual *electron-positron* pairs that effectively alter the distribution of charges and currents that generated the original electromagnetic field.

Valley of stability. The region on a chart of the *nuclides* where the majority of stable nuclides are found.

W^+. A positively charged *boson* with a mass about 87 times that of the *proton* that mediates the *weak force*.

W^-. The negatively charged counterpart to the W^+.

Wave. A disturbance that propagates from one point in a medium to other points without giving the medium as a whole any permanent displacement.

Weak coupling. The coupling of four *fermion* fields in the *weak interaction*, having a strength many orders of magnitude weaker than that of the *strong* or electromagnetic *interactions*.

Weak force. See *weak interaction*.

Weak interaction. One of the fundamental interactions among *elementary particles* responsible for beta decay of nuclei and for the decay of particles with lifetimes greater than about 10^{-10} second, such as *muons*, *K mesons*, and *lambda hyperons*; it is several orders of magnitude weaker than the strong and electromagnetic interactions.

X-ray astronomy. The study of x rays mainly from sources outside the solar system; it includes the study of *novae* and *supernovae* in the Milky Way Galaxy, together with extragalactic radio sources.

X-ray tomography. See *tomography*.

Z^0. A neutral *boson* with a mass about 100 times that of the *proton* that mediates the *weak force*.

Appendix A: Panel Members

PANEL ON ATOMIC, MOLECULAR, AND OPTICAL PHYSICS

DANIEL KLEPPNER, Massachusetts Institute of Technology, *Chairman*
C. LEWIS COCKE, JR., Kansas State University
ALEXANDER DALGARNO, Harvard-Smithsonian Center for Astrophysics
ROBERT W. FIELD, Massachusetts Institute of Technology
THEODOR W. HÄNSCH, Stanford University
NEAL F. LANE, University of Colorado at Colorado Springs
JOSEPH H. MACEK, University of Nebraska
FRANCIS M. PIPKIN, Harvard University
IVAN A. SELLIN, Oak Ridge National Laboratory

PANEL ON CONDENSED-MATTER PHYSICS

ALEXEI A. MARADUDIN, University of California, Irvine, *Chairman*
NEIL W. ASHCROFT, Cornell University
JOHN D. AXE, Brookhaven National Laboratory
PRAVEEN CHAUDHARI, IBM T.J. Watson Research Center
C. PETER FLYNN, University of Illinois
JERRY P. GOLLUB, Haverford College
BERTRAND I. HALPERIN, Harvard University

158 APPENDIX A

DAVID L. HUBER, University of Wisconsin
RICHARD M. MARTIN, Xerox Corporation
DOUGLAS L. MILLS, University of California, Irvine
ROBERT C. RICHARDSON, Cornell University
JOHN M. ROWELL, AT&T Bell Laboratories

PANEL ON ELEMENTARY-PARTICLE PHYSICS

MARTIN L. PERL, Stanford Linear Accelerator Center, *Chairman*
CHARLES BALTAY, Columbia University
MARTIN BREIDENBACH, Stanford Linear Accelerator Center
GERALD FEINBERG, Columbia University
HOWARD A. GORDON, Brookhaven National Laboratory
LAWRENCE W. JONES, University of Michigan
BOYCE D. MCDANIEL, Cornell University
FRANK S. MERRITT, The University of Chicago
ROBERT P. PALMER, Brookhaven National Laboratory
JAMES M. PATERSON, Stanford Linear Accelerator Center
JOHN PEOPLES, JR., Fermi National Accelerator Laboratory
CHRIS QUIGG, Fermi National Accelerator Laboratory
DAVID M. RITSON, Stanford University
DAVID N. SCHRAMM, The University of Chicago
A. J. STEWART SMITH, Princeton University
MARK W. STROVINK, University of California, Berkeley

PANEL ON GRAVITATION, COSMOLOGY, AND COSMIC-RAY PHYSICS

DAVID T. WILKINSON, Princeton University, *Chairman*
PETER L. BENDER, University of Colorado
DOUGLAS M. EARDLEY, University of California, Santa Barbara
THOMAS K. GAISSER, University of Delaware
JAMES B. HARTLE, University of California, Santa Barbara
MARTIN H. ISRAEL, Washington University
LAWRENCE W. JONES, University of Michigan
R. BRUCE PARTRIDGE, Haverford College
DAVID N. SCHRAMM, The University of Chicago
IRWIN I. SHAPIRO, Harvard-Smithsonian Center for Astrophysics
ROBERT F. C. VESSOT, Harvard-Smithsonian Center for Astrophysics
ROBERT V. WAGONER, Stanford University

PANEL ON NUCLEAR PHYSICS

JOSEPH CERNY, University of California, Berkeley, and Lawrence Berkeley Laboratory, *Chairman*
PAUL T. DEBEVEC, University of Illinois, Urbana
ROBERT A. EISENSTEIN, Carnegie-Mellon University
NOEMIE BENCZER KOLLER, Rutgers University
STEVEN E. KOONIN, California Institute of Technology
PETER D. MACD. PARKER, Yale University
R. G. HAMISH ROBERTSON, Los Alamos National Laboratory
STEVEN E. VIGDOR, Indiana University
JOHN D. WALECKA, Stanford University

PANEL ON THE PHYSICS OF PLASMAS AND FLUIDS

RONALD C. DAVIDSON, Massachusetts Institute of Technology, *Co-chairman*
JOHN M. DAWSON, University of California, Los Angeles, *Co-chairman*
GEORGE BEKEFI, Massachusetts Institute of Technology
ROY GOULD, California Institute of Technology
ABRAHAM HERTZBERG, University of Washington
CHARLES F. KENNEL, University of California, Los Angeles
LOUIS J. LANZEROTTI, AT&T Bell Laboratories
E. P. MUNTZ, University of Southern California
RICHARD F. POST, Lawrence Livermore National Laboratory
NORMAN ROSTOKER, University of California, Irvine
PAUL H. RUTHERFORD, Princeton University Plasma Physics Laboratory

PANEL ON SCIENTIFIC INTERFACES AND TECHNOLOGICAL APPLICATIONS

WATT W. WEBB, Cornell University, *Co-Chairman*
PAUL A. FLEURY, AT&T Bell Laboratories, *Co-Chairman*
TED G. BERLINCOURT, Office of Naval Research
AARON N. BLOCH, EXXON Research and Engineering
ROBERT A. BUHRMAN, Cornell University
PETER M. EISENBERGER, EXXON Research and Engineering
MITCHELL FEIGENBAUM, Cornell University
KENNETH A. JACKSON, AT&T Bell Laboratories
ANGELO A. LAMOLA, Polaroid Corporation

JAMES S. LANGER, University of California, Santa Barbara
ROBERT L. PARK, University of Maryland
WILLIAM H. PRESS, Harvard University
DONALD TURCOTTE, Cornell University
ALEXANDER ZUCKER, Oak Ridge National Laboratory

Index

A

Abbreviations and acronyms, 139-141
Advisory committees, 136
Alternating Gradient Synchrotron (AGS), 61
AMO physics, *see* Atomic, molecular, and optical physics
Artificially structured materials, 25, 26, 41
Astrophysical plasma, 67
Astrophysics, 7, 67
  vigorous space program in, 68
Asymptotic freedom, 19
Atomic, molecular, and optical (AMO) physics, 2, 13
  federal obligations for, 129, 132
  funding history for, 79
  panel on, 157
  progress in, 28-31
Atoms, 13, 18

B

BCX (Burning Core Experiment), 56
Big bang, 14, 16, 37-38
Biophysics, 8
  interface of physics with, 40
Black holes, 37
BNL (Brookhaven National Laboratory), 59
Bosons, 24
Bottom quark, 18, 19
Breathing mode, 23
Brookhaven National Laboratory (BNL), 59
Burning Core Experiment (BCX), 56

C

CARS (coherent anti-Stokes Raman scattering), 28
CDF (Collider Detector) at Fermilab, 60
CEBAF (Continuous Electron Beam Accelerator Facility), 56, 61, 62
CERN (Conseil Européen de Recherche Nucléaire), 79, 86
CESR (Cornell Electron-Positron Storage Ring), 54, 61
Chaotic motion, 17
Charm quark, 18, 19
Chemistry, interface of physics with, 39-40
Chromodynamics, quantum, 19

162 INDEX

Coherent anti-Stokes Raman scattering (CARS), 28
Collaborative exchange programs, 86-87
Collider Detector at Fermilab (CDF), 60
Combustion, 42
Committee on Education of the American Physical Society, 47-48
Computers, 72-73
Condensed-matter physics, 2, 13
 federal obligations for, 129, 132
 funding history for, 79-80
 major advances in, 84-85
 panel on, 157-158
 progress in, 24-28
 recommended program for, 63-65
Conseil Européen de Recherche Nucléaire (CERN), 79, 85-86
Construction facilities, 134-135
Continuous Electron Beam Accelerator Facility (CEBAF), 56, 61, 62
Cooperation, international, 90
Cornell Electron-Positron Storage Ring (CESR), 54, 61
Cosmic rays, 15
 federal obligations for research in, 130, 133
 ground-based, 69
 highest-energy, 39
 long-duration experiments with, 69
 panel on, 158
 progress in study of, 38-39
Cosmological principle, 37
Cosmology, 3, 7, 14
 federal obligations for, 130, 133
 panel on, 158
 progress in, 37-38
 vigorous space program in, 68
CPI-W index, 119

D

Data bases, 73
Degree holders, physics, retention of, 98-99
Degree production, physics, 94
Demand projections, physicists, 103-108
 in academe, 103-105
 in 4-year colleges, 105-106, 108
 in industrial and other nonacademic sectors, 106-108, 109
 in universities, 105, 106, 107

Demand-supply balance, physicist, 112-114
Department of Defense (DOD), 71
Department of Defense-University Instrumentation Program, 51
Disordered materials, 27
Doctoral degrees, physics, 93
Doctorates, physics, by subfield, 98
DOD (Department of Defense), 71
Down quark, 18, 19

E

Economic growth, 10
Education, 2, 45
 Committee on, of the American Physical Society, 47-48
 of foreign physicists in U.S., 87-90
 of foreign students enrolled in graduate physics programs in U.S., 89
 graduate, 48-49
 of next generation of physicists, 46-49
 primary, 46-47
 secondary, 46-47, 92
 supply of physicists and, 91-114
 undergraduate, 47-48
Electromagnetic
 force, 20-21
 theory, 7
 traps, 28-29
Electron-plasma oscillation, 14
Electron-positron
 collider, 60-61
 pair, 31
Electrons, 19
Electroweak interactions, 11-12
Elementary-particle physics, 2, 11-12
 avoiding duplication of research in, 86
 construction projects in, 134
 funding history for, 79-81
 important achievements in, 84
 major experimental advances in, 84
 NSF and DOE funding for, 125, 126, 129, 131
 panel on, 158
 progress in, 18-21
 recommended program for, 58-61
Employment, physics, patterns of, 101-102
Endoscopes, fiber-optic, 43

Energy, environment and, 42
Enrollments, 95-98
  U.S. and foreign composition, 95, 96
  women and minorities, 95-98
Environment, energy and, 42
EXAFS (extended x-ray absorption fine structure), 63, 64
Exchange programs, collaborative, 86-87
Expenditures for scientific research, 76-81
Extended x-ray absorption fine structure (EXAFS), 63, 64

F

Federal funding for research, 119-135
  for applied research, 122, 123
  for basic research, 120, 121
  for subfields of physics, 129-130
Fermi National Accelerator Laboratory (FNAL), 59, 61
Fiber-optic endoscopes, 43
Fluctuations, 27
Fluid physics, 2
  federal obligations for, 130, 133
  panel on, 159
  progress in, 35
Fluids, nonideal, 35
Fly's Eye facility, 69
FNAL (Fermi National Accelerator Laboratory), 59, 61
Forces, four fundamental, 20-21
Free-electron laser, 14
Funding process, research, 45-46
Fusion
  magnetic, 32-33
  physics, construction projects in, 134
  plasma, 66
Fusion-oriented research, 86

G

G (Newton's gravitational constant), 36
Galactic plasma, 14
Gamma rays, 15
  discrete sources of, 39
General relativity, 35-37
Generation model, 20
Geophysics, physics interface with, 40-41

Glasses, 27
  spin, 27-28
Glossary of physical terms, 142-156
Gluons, 2, 12
GNP (gross national product), 76-81
Graduate
  education, 48-49
  students, physics, 92-94
Grand Unification Theory, 37-38
Gravitation physics, 3, 15
  federal obligations for, 130, 133
  panel on, 158
  progress in, 35-37
Gravitational
  constant G, Newton's, 36
  force, 20-21
  radiation, 36, 68
  redshift, 36
Gravity Probe B Satellite, 56
Gravity-Wave Detector, 56
Gross national product (GNP), 76-81
Gyrotron, 32

H

Hadrons, 18
Hall effect, quantized, 25
HEP indices, 119
High Energy Physics Advisory Panel (HEPAP), 136
Hypernuclei, 12, 23

I

Industry
  expenditures on basic research, 124
  physics and, 43
  research in, 117-118
  role of, in basic research, 71
Inertial fusion research, 66-67
Information, free flow of, 87
Instrumentation, 53
  advanced, 50
  scale and costs of, 85-86
Interacting boson model, 24
Interfaces of physics, 15
  with biophysics, 40
  with chemistry, 39-40
  with geophysics, 40-41
  with materials science, 41

164   INDEX

panel on, 159-160
progress in, 39-41
International
    aspects of physics, 75-90
    competition and cooperation, 85-87, 90
    enterprise, physics as, 71
    position of U.S. physics, 5
Internationalization, increased, of physics community, 85
Ion traps, 29, 30

J

Joint European Torus (JET), 33

L

Lamb shift, 29
Large-amplitude space-charge waves, 31-32
Laser
    interferometer, 56
    spectroscopy, 28
Lasers, 2, 13, 17, 42-43
    free-electron, 14
Lawson confinement parameter, 33
Leptonic atoms, 29
Leptons, 2, 11-12
    as elementary particles, 18-20
Localization, 27

M

Magnetic
    fields, high, 65
    fusion, 32-33
    Fusion Advisory Committee (MFAC), 136
    fusion research, 65-66
    resonance imaging (MRI), 8-9
Magnetic-field reconnection, 32
Magnetospheres, 14, 34
Manpower, see Physicists
Mansfield Amendment, 130
Mass, missing, in universe, 17
Materials science, physics interface with, 41
Medicine, physics and, 42-43
Meson exchange currents, 62
Metalorganic chemical vapor deposition (MOCVD), 26
MFAC (Magnetic Fusion Advisory Committee), 136
Microcomputers, 8
Microelectronics, 41, 43
Mirror confinement systems, 33
Mission agencies, role of, in basic research, 71
Mobility, outward, physicists and, 99
MOCVD (metallorganic chemical vapor deposition), 26
Molecular
    channels, single, 40
    physics, see Atomic, molecular, and optical physics
Mortality schedules, TIAA-CREF, 104
MRI (magnetic resonance imaging), 8-9
Multidisciplinary facilities, 135
Muon neutrino, 19, 20
Muonium, 29

N

National
    laboratories, 116-117
    Science Foundation (NSF), 72-73, 116
    security, 43
    Synchrotron Light Source, 57
Neutral currents, 20-21
Neutrino astronomy, 69
Neutrinos, 17, 19-20
Neutron facilities, 64-65
Neutrons, 12
Newton's gravitational constant G, 36
Nobel Prize in Physics, 81, 82
    recipients of, 83
Nonideal fluids, 35
NSAC (Nuclear Science Advisory Committee), 136
NSF (National Science Foundation), 72-73, 116
Nuclear
    physics, 2, 12
        construction projects in, 134
        NSF and DOE funding for, 127, 128, 129, 131
        panel on, 159
        progress in, 21-24
        recommended program for, 61-63
    Science Advisory Committee (NSAC), 136

stability, 23
Nucleons, 2, 12
Nucleus, 18
  model of, 21-24

O

Optical physics, *see* Atomic, molecular, and optical physics
Optical-frequency counting methods, 13
Order-disorder transitions, 39-40
Organic conductors, 40
Outward mobility, physicists and, 99

P

Packard report, 116, 117
Panel members, 157-160
Particle accelerators, 11, 61-63
Particle physics, *see* Elementary-particle physics
Particle-Beam Fusion Accelerator (PBFA), 55
Particle-trap techniques, 13
PBFA (Particle-Beam Fusion Accelerator), 55
Phase transitions, 2, 25, 27
Ph.D. production, physics, 109-114
Photoemission studies, 24
Physical terms, glossary of, 142-156
Physicists
  aging community of, 99-101
  demand projections, *see* Demand projections, physicists
  demand-supply balance, 112, 114
  educating, 3, 45; *see also* Education, next generation of, 46-49
  European, 8
  excellence of, 69-70
  female, 4, 70
  foreign-born, 70
    education of, in U.S., 87-90
  freedom for, 87
  future demand for, 69-70
  producing trained young, 92-95
  production of, 4
  projections of demand and supply of, 102-113
  retirement rate, 91
  role of, 9
  supply projections, 108-113
Physics, 1, 6
  applications of, 15, 16
    panel on, 159-160
    progress in, 42-43
  degree holders, retention of, 98-99
  degree production, 94
  discoveries in, 2-3
  doctoral degrees, 93
  doctorates by subfield, 98
  education in, *see* Education
  employment, patterns of, 101-102
  enrollments in, *see* Enrollments
  excellence in, 44-45
  graduate, foreign students in, 89
  graduate students, 92-94
  impact of, 3
  increased internationalization in, 85
  industry and, 43
  interfaces of, *see* Interfaces of physics
  international aspects of, *see* International *entries*
  large facilities and major programs in, 3-4, 53-69
  maintaining excellence in, 44-73
  medicine and, 42-43
  nation and, 2-3
  Nobel Prize in, *see* Nobel Prize in Physics
  organization and decision making in, 135-137
  organization and support of, 115-137
  Ph.D. production, 109-114
  progress in, 11-43
  proposed large construction projects in, 56
  research in, *see* Research
  research publications in, 82
  role of, in U.S., 8-10
  scientific manpower in, *see* Physicists
  small-group, 49-53
  society and, 6-10
  trends in specific areas of, 79-81
  unity of, 15-18
  U.S. leadership role in, 91
Plasma physics, 2, 4
  construction projects, 134
  federal obligations for, 130, 133
  panel on, 159
  progress in, 31-34

recommended program for, 65-67
Plasmas, 14, 31
  astrophysical, 67
  fusion, 66
  galactic, 14
  quark-gluon, 12, 22, 24, 62
  space, 34, 67
Policy issues connected with physics, 70-71
Positive energy theorem, 37
Positronium, 29
Primary education, physics and, 46-47
Proton decay, 21, 84
Protons, 12
Publications, number of, 82
Pulsars, 36-37
Pulsed spallation sources, 64

Q

Quantized Hall effect, 25
Quantum
  chromodynamics, 19
  mechanics, 7
Quark-gluon
  model, 16
  plasma, 12, 22, 24, 62
Quarks, 2, 11-12
  confinement of, 23
  as elementary particles, 18-20

R

Relativistic Nuclear Collider (RNC), 56, 61, 62
Relativity
  general, 35-37
  gyroscope experiment, 68
Renormalization group approach, 27
Research, 1
  academic, 3
  avoiding duplication in, 86
  basic
    role of industry and mission agencies in, 71
    U.S. position in, 81-85
  collaborative, 86-87
  complementary roles in, 118
  cutting-edge, 53, 57
  diversity of institutions for, 115-118
  funding process, 45-46
  funding support for, 119-135
  fusion-oriented, 86
  in industry, 117-118
  location of, 75
  main types of large facilities used in, 56
  publications in physics, 82
  quality of, 82
  scientific, expenditures for, 76-81
  in small groups, 3, 49-53
  supporting, 4
  trends in expenditures in, 79-81
  universities and, 3, 48-49, 117
RNC (Relativistic Nuclear Collider), 56, 61, 62

S

Scientists, *see* Physicists
Secondary education
  crisis in, 92
  physics and, 46-47
Security, national, 43
Seismology, 40-41
  underwater, 41
Semiconductors, 7-8
Shell-model orbits, 23
Simulation physics, 72
SLAC (Stanford Linear Accelerator Center), 59, 60
Small-group physics, 3, 49-53
Society, physics and, 6-10
Solar wind, 14
Solitons, 32
Space
  plasma, 34, 67
  program, vigorous, in astrophysics, 68
Spectroscopy, 17
  advances in, 39
Spin glass, 27-28
SSC (Superconducting Super Collider), 56, 58-59
Stanford Linear Accelerator Center (SLAC), 59, 60
Strange quark, 18, 19
Strong force, 20-21
Summary, 1-5
Supercomputers, 72
Superconducting Super Collider (SSC), 56, 58-59

Superconductivity, 13
Superconductors, 18
Supersymmetries, 24
Supply projections, physicists, 108-112
Surfaces, understanding of, 24-25
Synchrotron
  Facility, 6-GeV, 56
  radiation, 63
  radiation facilities, 17-18, 63-64

T

Tandem mirror concept, 33
Tau
  lepton, 19, 20
  neutrino, 19, 20
Tevatron, 60, 61
TFTR (Tokamak Fusion Test Reactor), 33, 54, 66
Thermonuclear fusion, 65-67
TIAA-CREF mortality schedules, 104
Tokamak
  devices, 32-33
  Fusion Test Reactor (TFTR), 33, 54, 66
Top quark, 18, 19
Transistors, 7-8

Tunneling microscope, 24-25
Turbulence, 17, 35
  fundamental theory of, 41

U

Ultraviolet Radiation Synchrotron, 57
Undergraduate education, 47-48
Unification of forces of nature, 20-21
Universe
  isotropic, 37
  missing mass in, 17
Universities
  demand scenarios for physicists in, 105, 106, 107
  research and, 3, 48-49, 117
  small-group research in, 3, 51-53
Up quark, 18, 19

W

W particles, 20-21
Weak force, 20-21

Z

Z particle, 20-21